Playing with Numbers: Challenges of Logic for Ingenious Children

MARILEDYS TOVAR

WELCOME TO
MATHEMAGIC WORLD!

Hello little math explorers! I present to you "Playing with Numbers: Logic Challenges for Ingenious Children," a book from the Mathemagic World series, designed to awaken curiosity and love for mathematics in children ages 6 to 8.

Playing with Numbers: Logic Challenges for Ingenious Kids has been carefully designed to stimulate children's minds and make mathematical learning an exciting experience. You'll find fun activities like mazes, puzzles, comparisons, and creative challenges that encourage curiosity and problem solving; in addition to adventures that will bring into play ingenuity and creativity to solve numerical puzzles that will facilitate the learning and mastery of addition and subtraction.

Numbers will be your guide in this Mathemagic World! Are you ready to accept the challenge?

"Logic Challenges for Ingenious Kids" awaits you to explore, learn and have fun with the wonders of mathematics!

Let the adventure begin!

MARILEDYS TOVAR

I BELONG TO:

hello

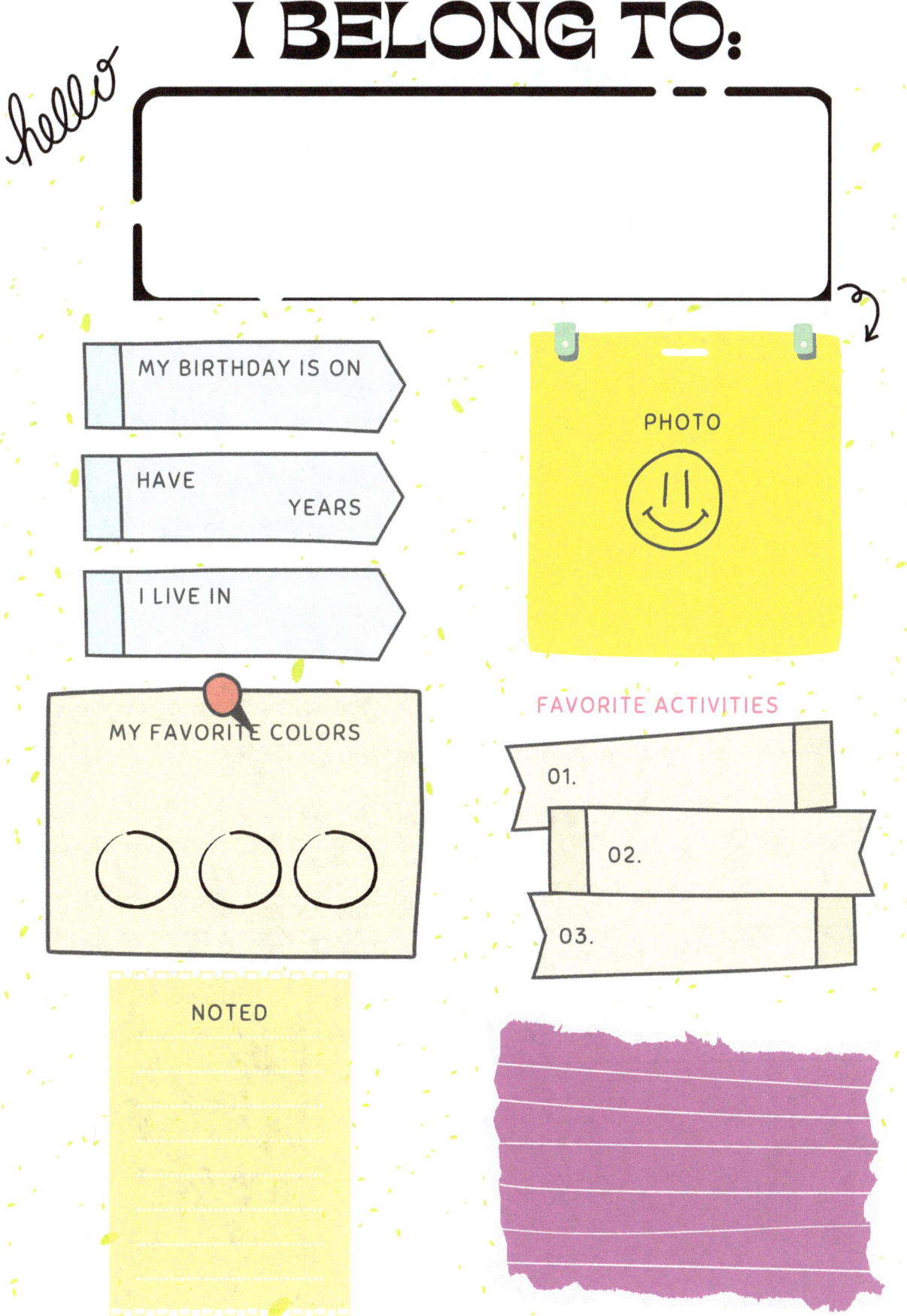

MY BIRTHDAY IS ON

HAVE   YEARS

I LIVE IN

PHOTO

MY FAVORITE COLORS

FAVORITE ACTIVITIES

01.

02.

03.

NOTED

Scan to receive free information
and resources

You can also write to
mariledys@educkidsonline.com

WE GO TO SCHOOL

Draw a line and help the children get to school.

DAILY RHYTHM

Identify the routine, draw the clock hands and write the time

1.		I wake up at
2.		I take a shower at
3.		Breakfast at
4.		I am going to the school at
5.		Lunch at
6.		I take a nap at
7.		I do my homework at
8.		I watch television at
9.		I eat at
10.		I go to bed at

WHAT TIME IS IT?

Write the time shown on the clocks

DISCOVER YOUR TIME!

Select the correct answer in each statement

If you must sleep for 9 hours at night and wake up at 7:00 in the morning, what time should you go to bed?

9:30 **11:00** **10:00** **12:00**

If you start playing at 4:45 in the afternoon and play for 2 hours, what time do you finish playing?

7:30 **6:45** **5:45** **8:00**

If dinner is ready at 7:15 PM and it takes you 30 minutes to eat, what time will you finish dinner?

7:45 **6:45** **4:45** **7:00**

If you start doing your homework at 5:45 PM and work for 30 minutes, what time will you finish?

6:06 **7:45** **5:45** **6:15**

HOW LONG HAS IT BEEN?

Read and calculate elapsed time

It takes Paula 1 hour and 20 minutes from her house to school. If you leave at 7:10, what time do you arrive at school?

[ ] hours and [ ] minutes

José takes 35 minutes walking from his house to the park. What time should you leave the house to arrive at the park at 9:40?

[ ] hours and [ ] minutes

Caro is counting the time until she leaves school. If it is 12:20 and he leaves school at 4:30 p.m. How much time left?

[ ] hours and [ ] minutes

I was in the library from 8:30 to 10:30. How many hours and minutes was I in the library?

[ ] hours and [ ] minutes

On Tuesday I went to swimming lessons from 9:45 to 11:30. How many hours and minutes was I in class?

[ ] hours and [ ] minutes

On Saturday I watched TV for 25 minutes, on Sunday for 1 hour and 10 minutes. How many hours and minutes of television did I watch in total?

[ ] hours and [ ] minutes

TIMELINE

Observe and respond

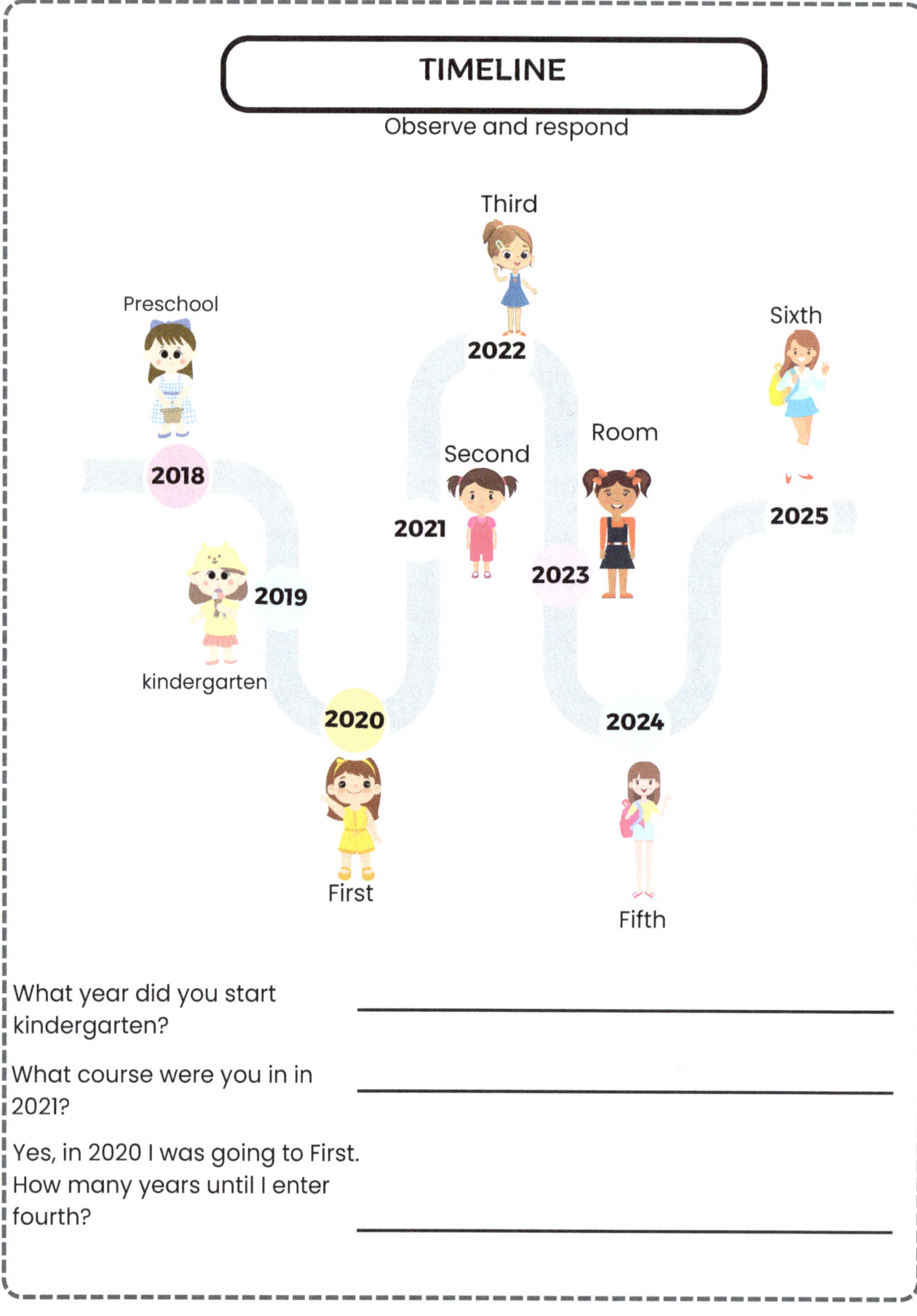

What year did you start kindergarten?

_____

What course were you in in 2021?

_____

Yes, in 2020 I was going to First. How many years until I enter fourth?

_____

TIMELINE

Create your timeline

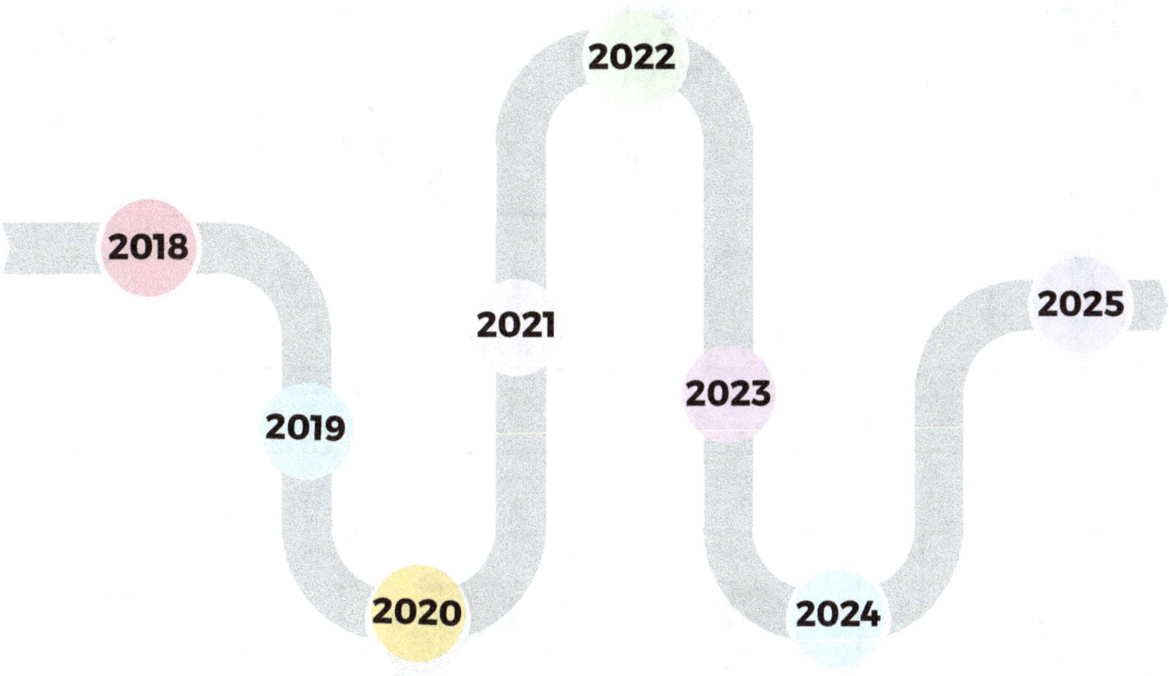

2018

2019

2020

2021

2022

2023

2024

2025

LOGIC SERIES WITH FIGURES

Check the correct option

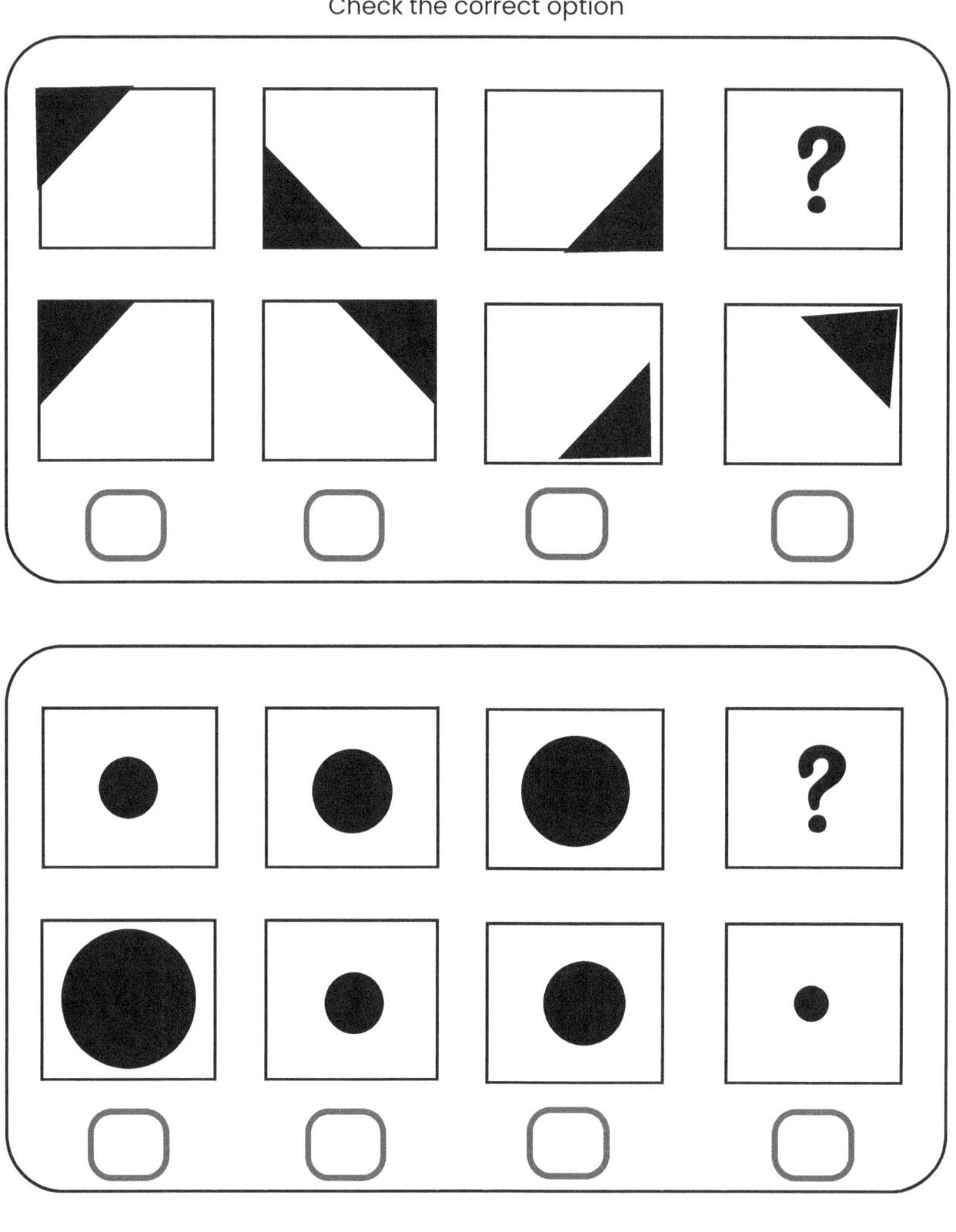

LOGIC SERIES WITH FIGURES

Check the correct option

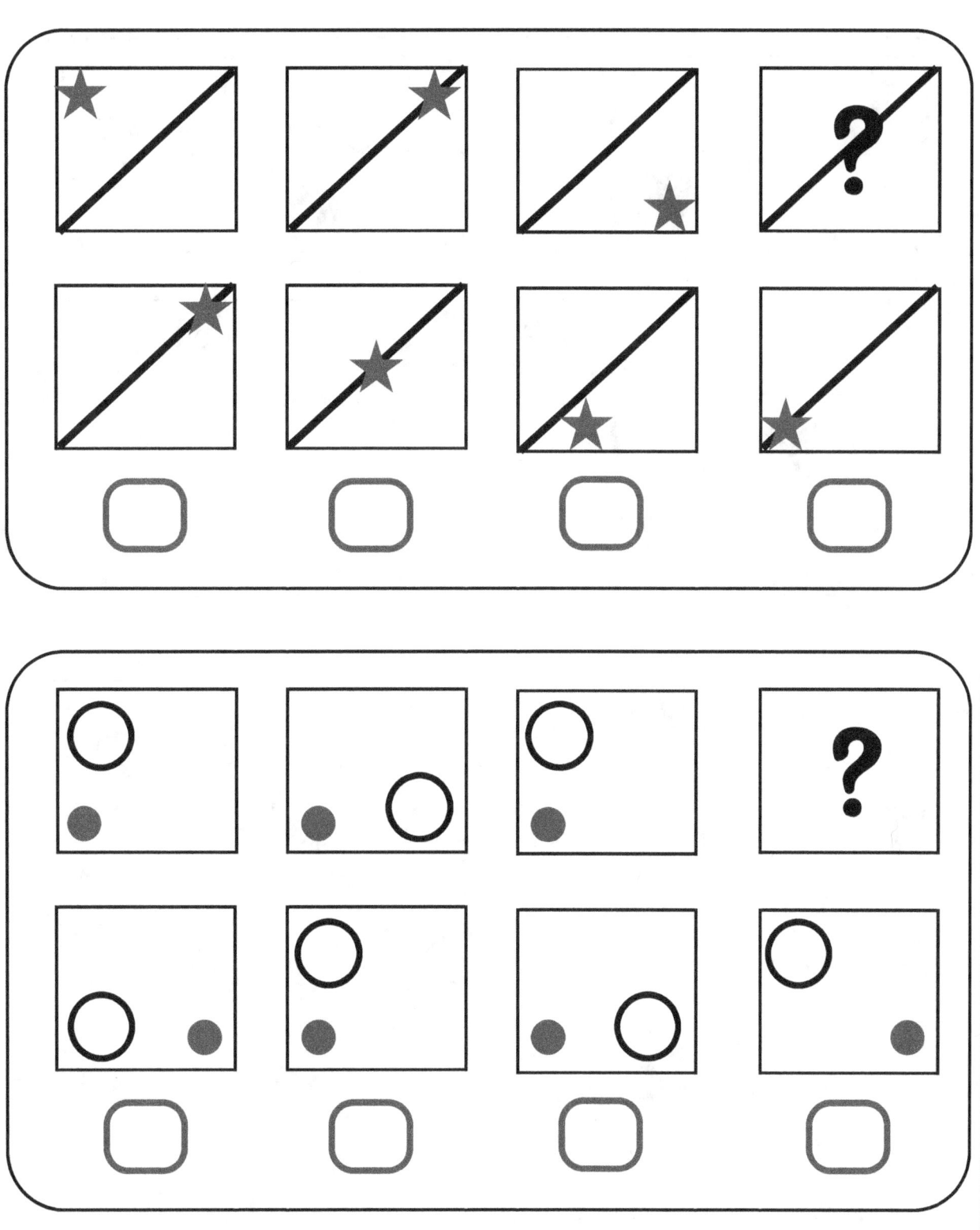

LOGIC SERIES WITH FIGURES

Check the correct option

LOGIC SERIES WITH FIGURES

Check the correct option

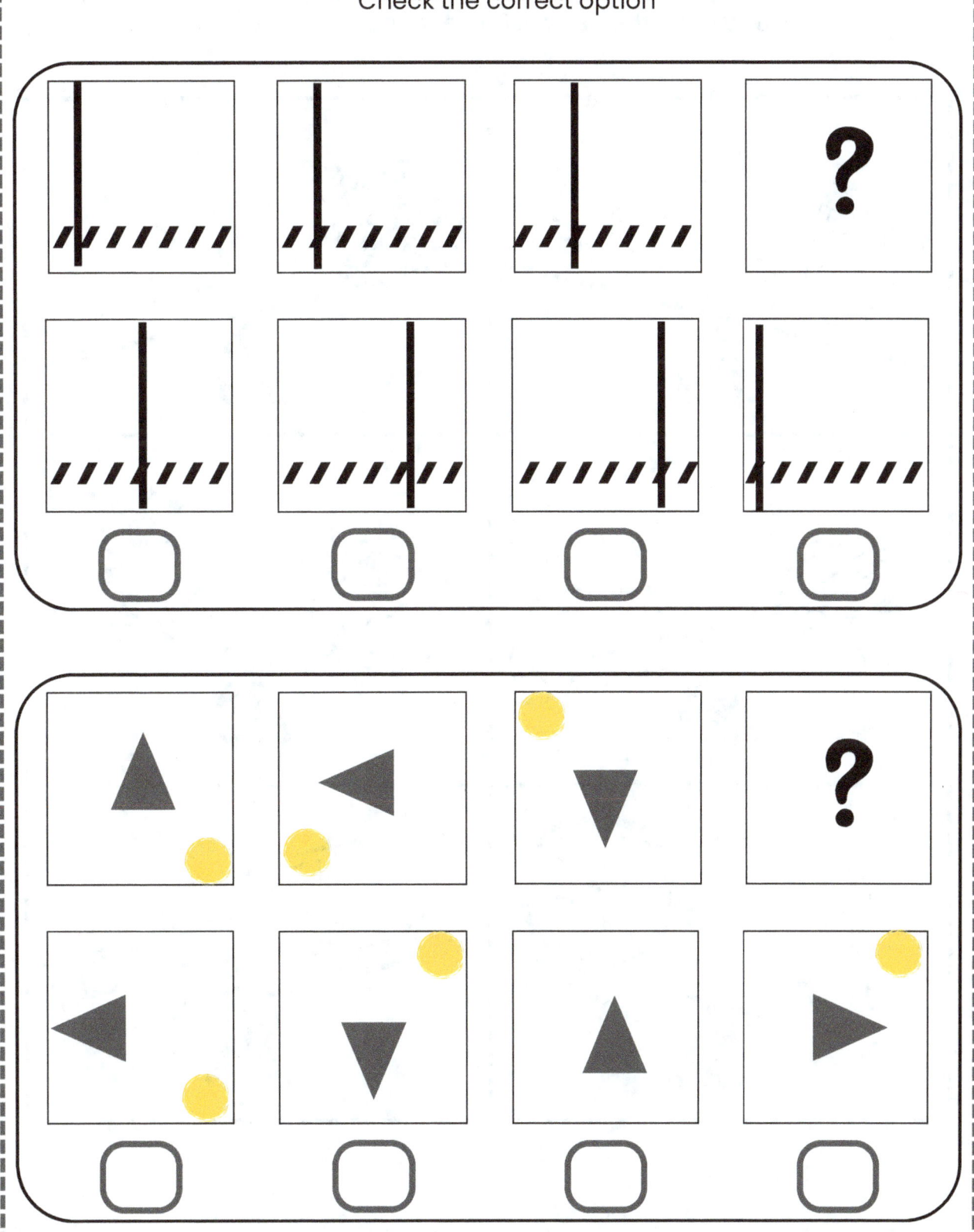

IN THE SCALES!

Identify and color the heaviest object and circle the lightest one.

IN THE SCALES!

Look at the image of each number. Fill in the blank with: Heavier than, lighter than or as heavy as

1. The cat_____ the hedgehog.

2. The owl is _____ than the raccoon

3. The squirrel is_____ the rabbit

FIND THE WAY

Guide the child to the car. Color

FIND THE WAY

Help the astronaut

FIND THE WAY

Help the pilot get to his plane

FIND THE WAY

Find the safest way for your car to get to work

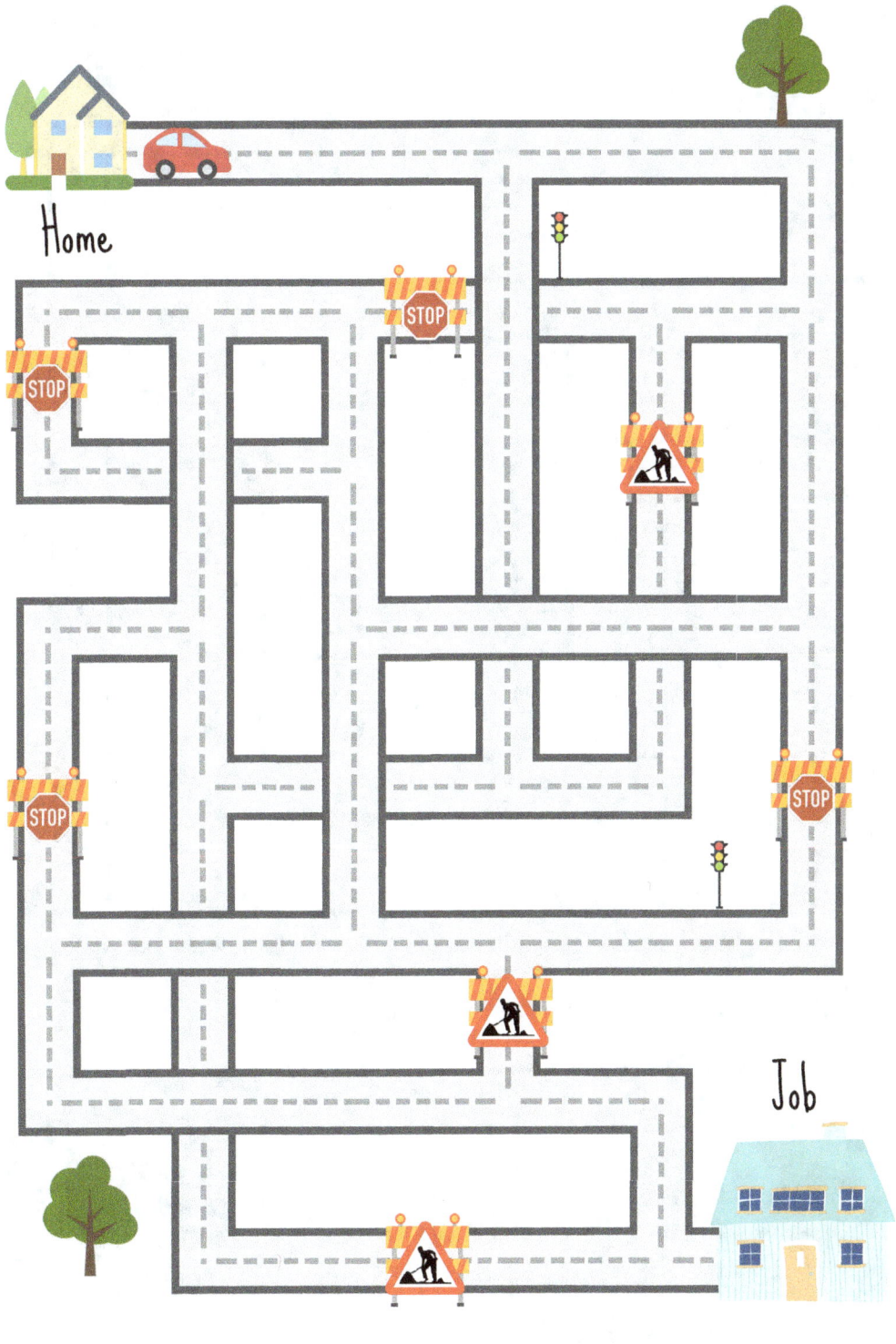

Home

Job

FIND THE WAY

Find the safest way for the child to get to the park

PUT THE PIECES TOGETHER!

Add the figures of each square in the final table

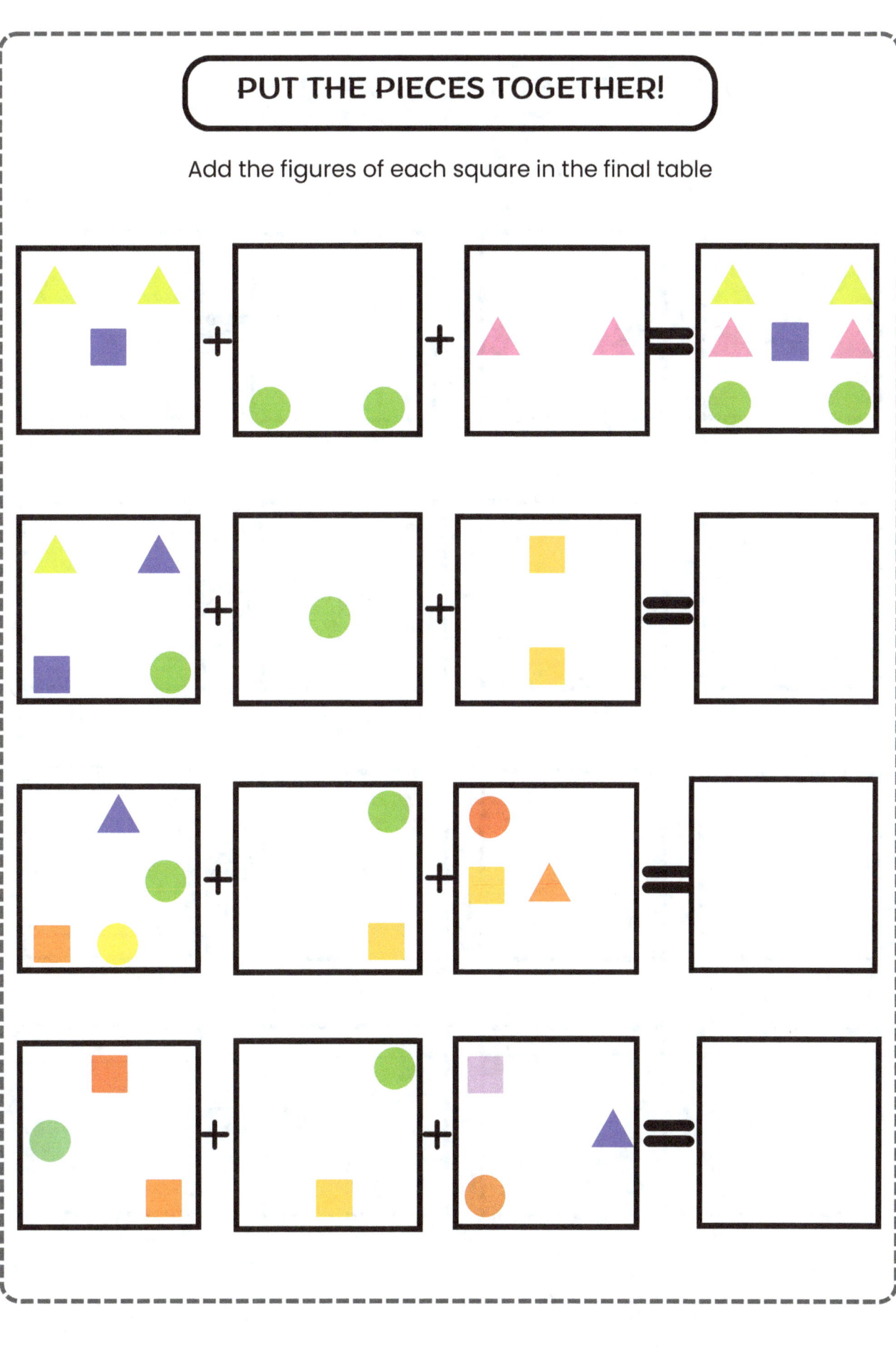

PUT THE PIECES TOGETHER!

Add the figures of each square in the final table

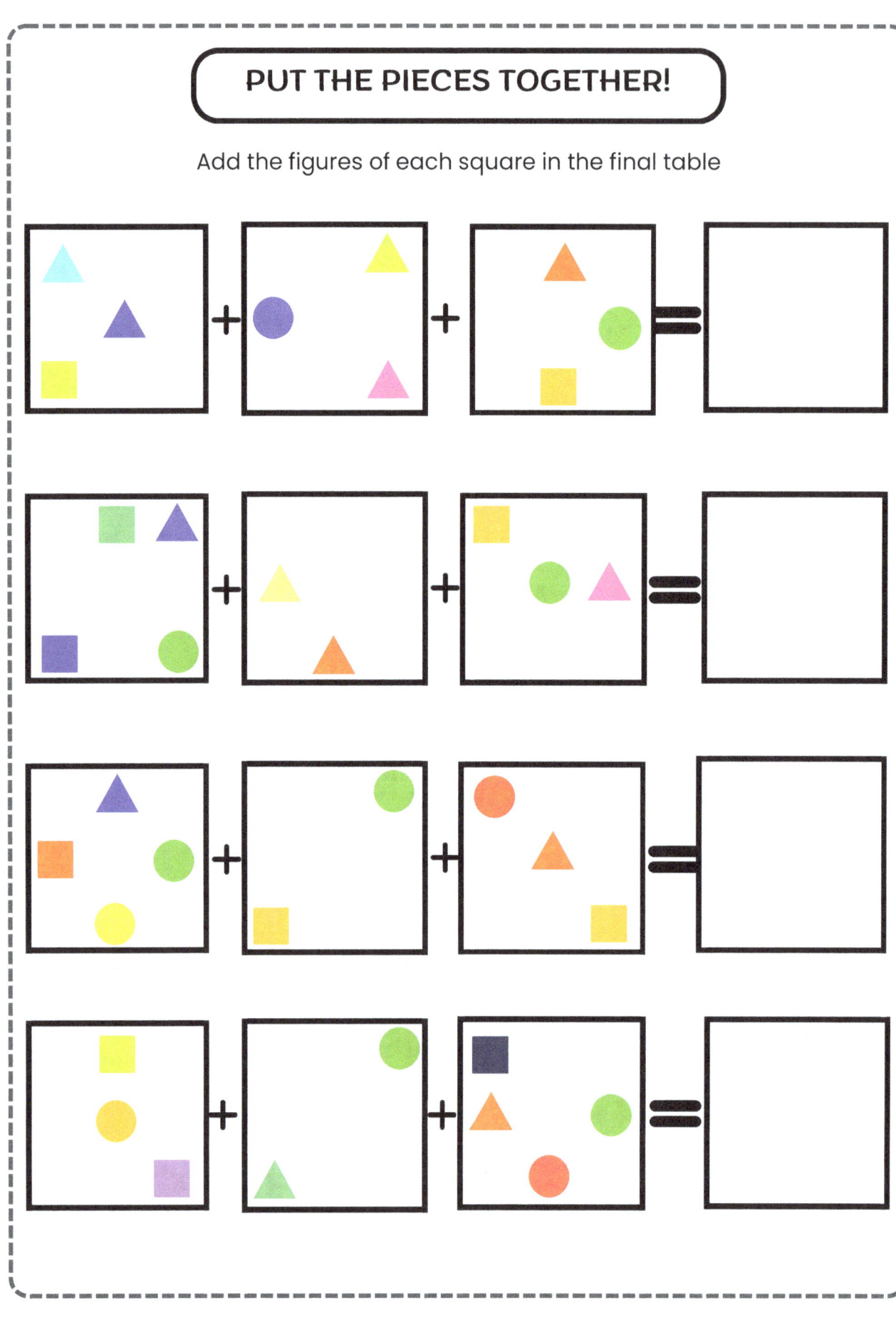

ENIGMATIC CLUES

Read, think, reason and respond

Of the 4 cyclists, it is known that Pepe has arrived immediately after Luís and María has arrived in the middle of Carla and Pepe. What is the order of arrival?

1

2

3

4

You have 4 cards with images of means of transportation: Helicopter, train, bus, plane. The helicopter card is to the left of the bus card, the plane card is to the right of the train. What is the order of the cards?

Laura has a red hat, a blue one and a green one. If you place the red hat under the blue and the blue under the green, what is the color of the hat on top? (Color)

ENIGMATIC CLUES

Read, think, reason and respond

Three brothers, Tom, Tim and Tony, are of consecutive ages. The sum of their ages is 27 years. We know that Tom is 8 years old. How old are Tim and Tony?

Tom____ years   Tim____ years   Tony____ years

You have four cards numbered 1 through 4. Card 2 is to the left of card 3, and card 4 is to the right of card 1. What is the correct order of the cards?

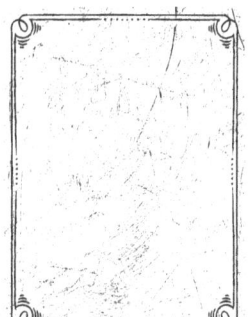   

In a row of 4 houses, Matías lives next to Lucía but not next to Carla. If Matías does not live next door to Jorge, who are Jorge's next-door neighbors?

LET'S COUNT TO 50

Fill in the missing numbers to reach 50

									10
									20
									30
									40
									50

PYRAMIDS OF NUMBERS

Order the pyramid from smallest to largest with the indicated numbers

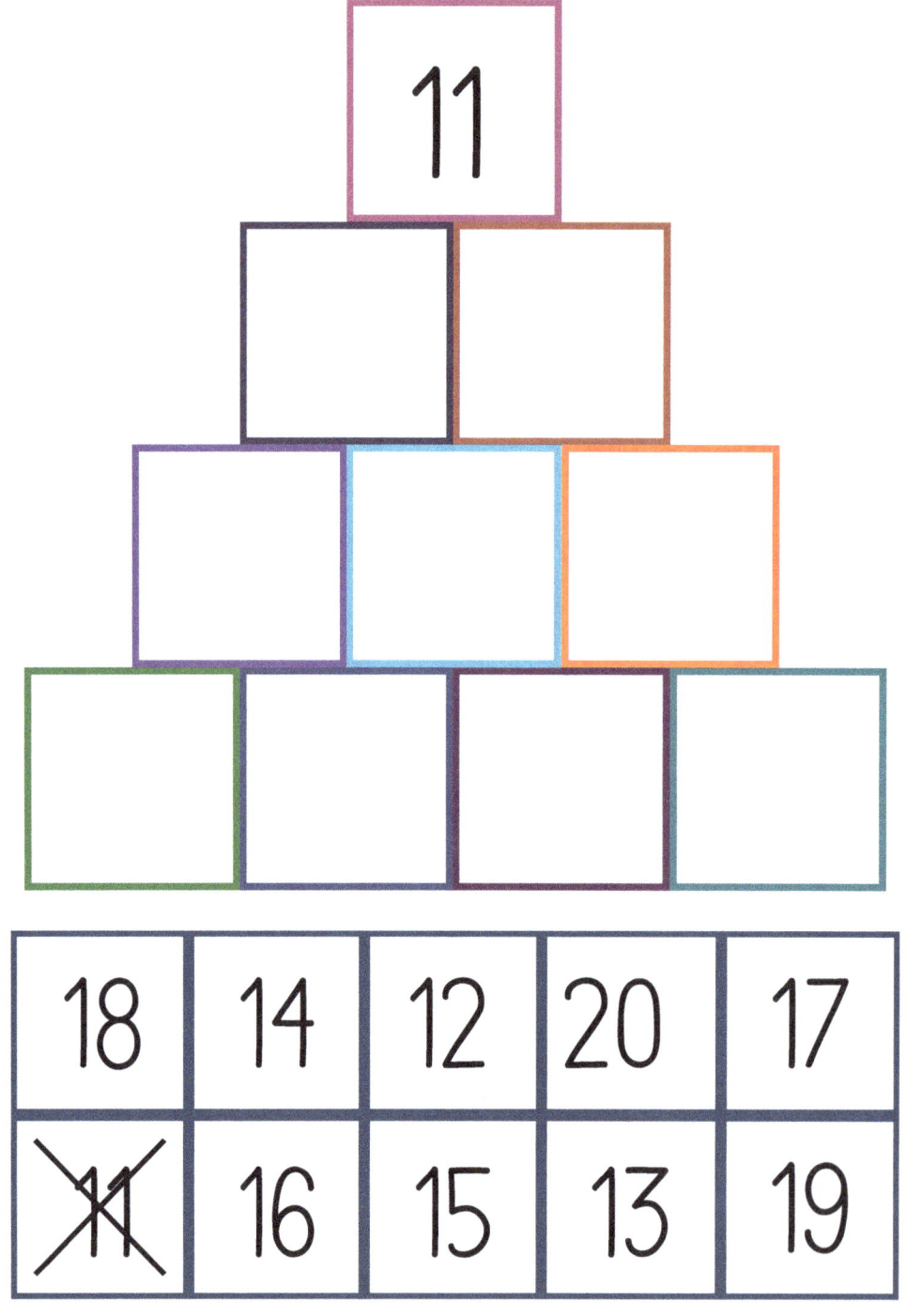

PYRAMIDS OF NUMBERS

Order the pyramid from smallest to largest with the indicated numbers

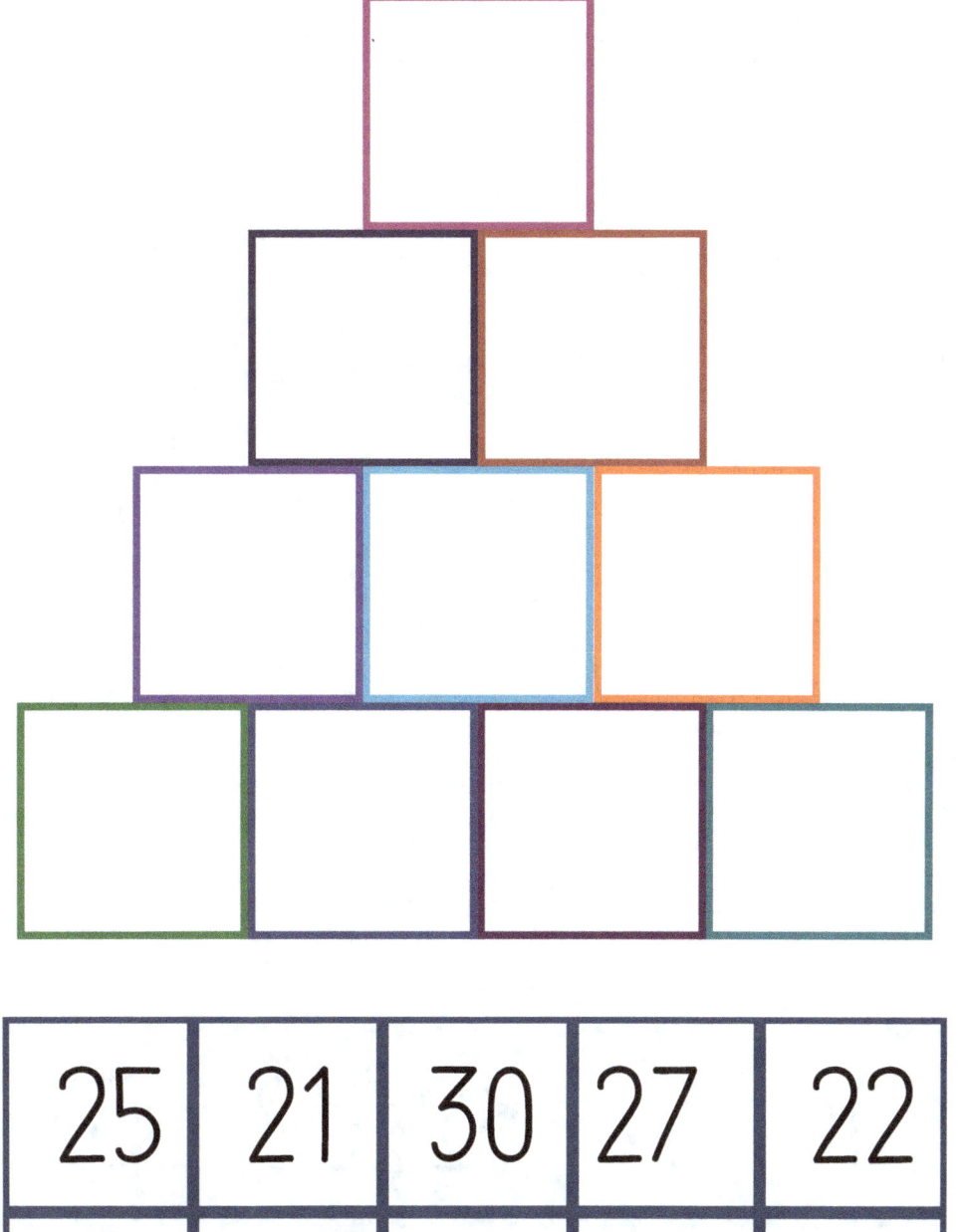

25	21	30	27	22
26	23	28	29	24

PYRAMIDS OF NUMBERS

Order the pyramid from smallest to largest with the indicated numbers

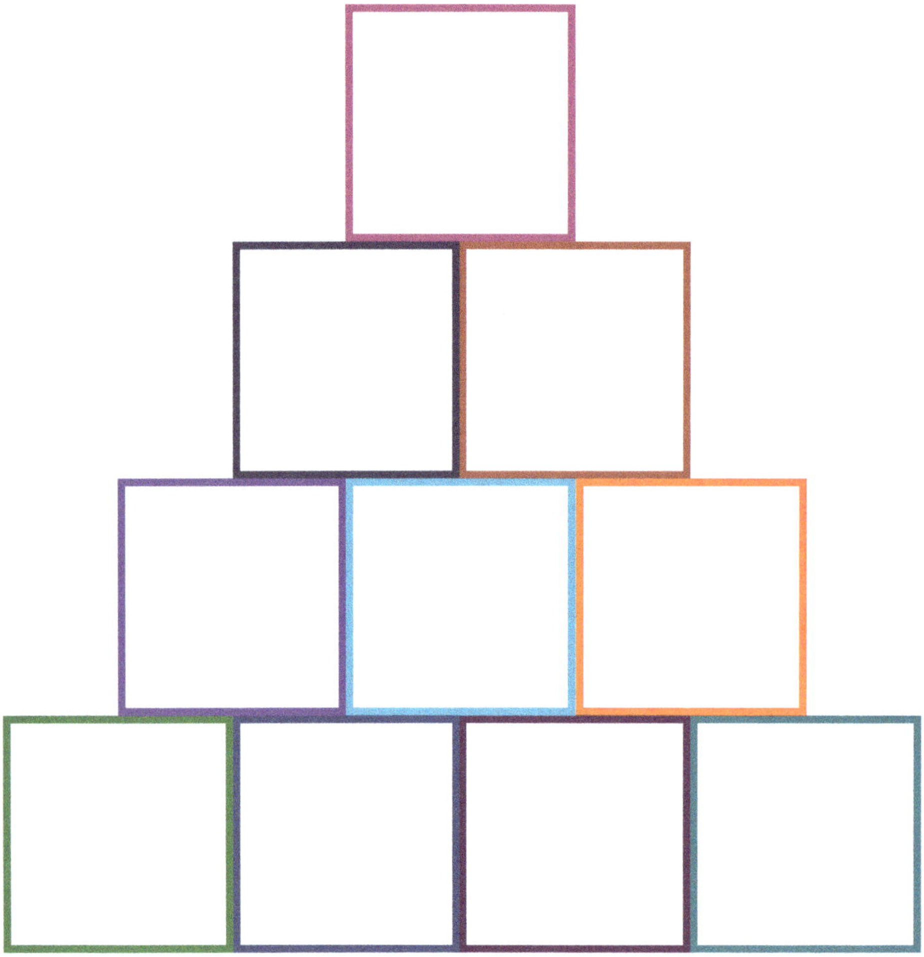

40	37	32	38	35
31	39	36	33	34

PYRAMIDS OF NUMBERS

Order the pyramid from smallest to largest with the indicated numbers

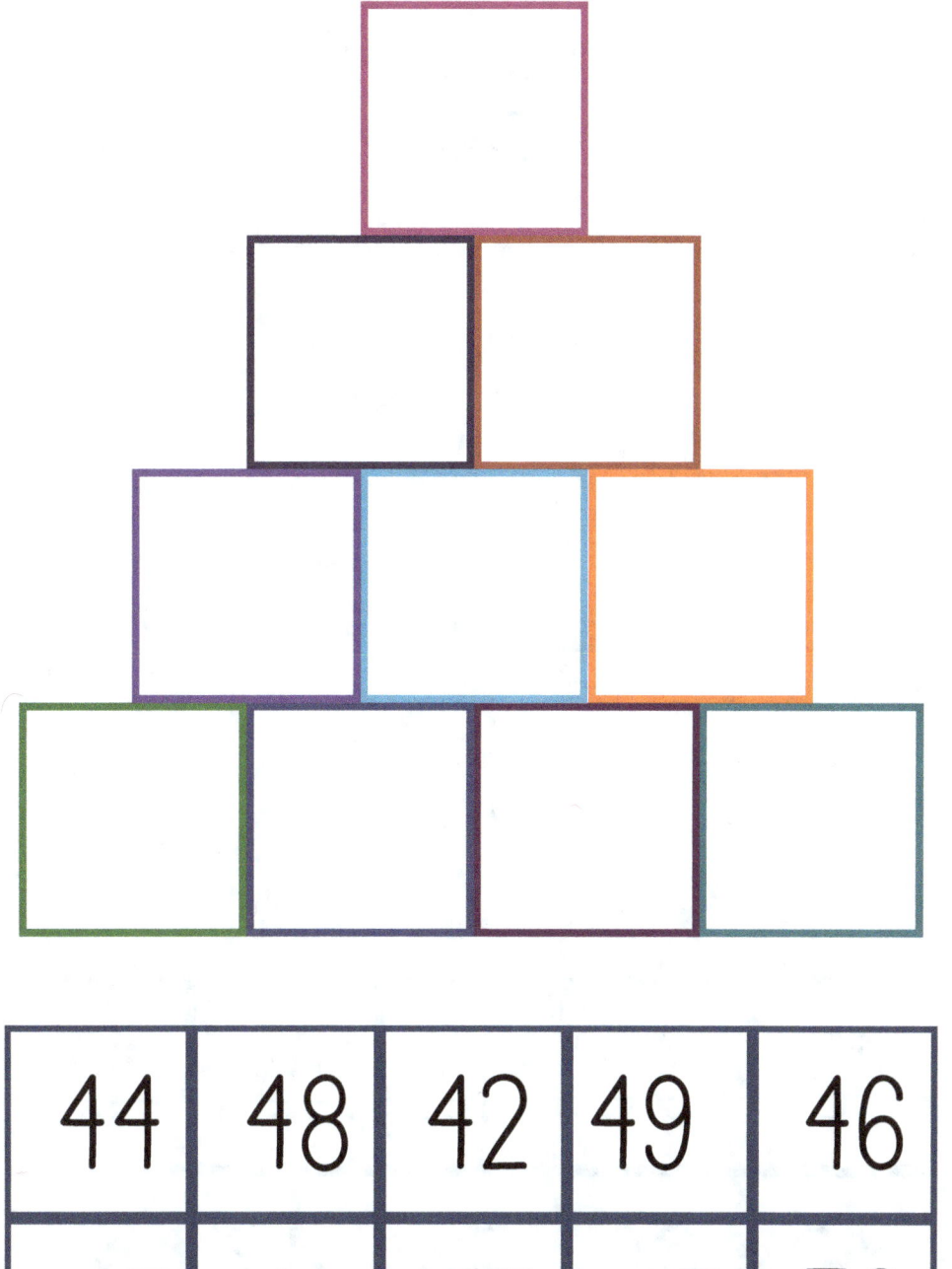

44	48	42	49	46
45	41	43	47	50

PYRAMIDS OF NUMBERS

Order the pyramid from largest to smallest with the indicated numbers

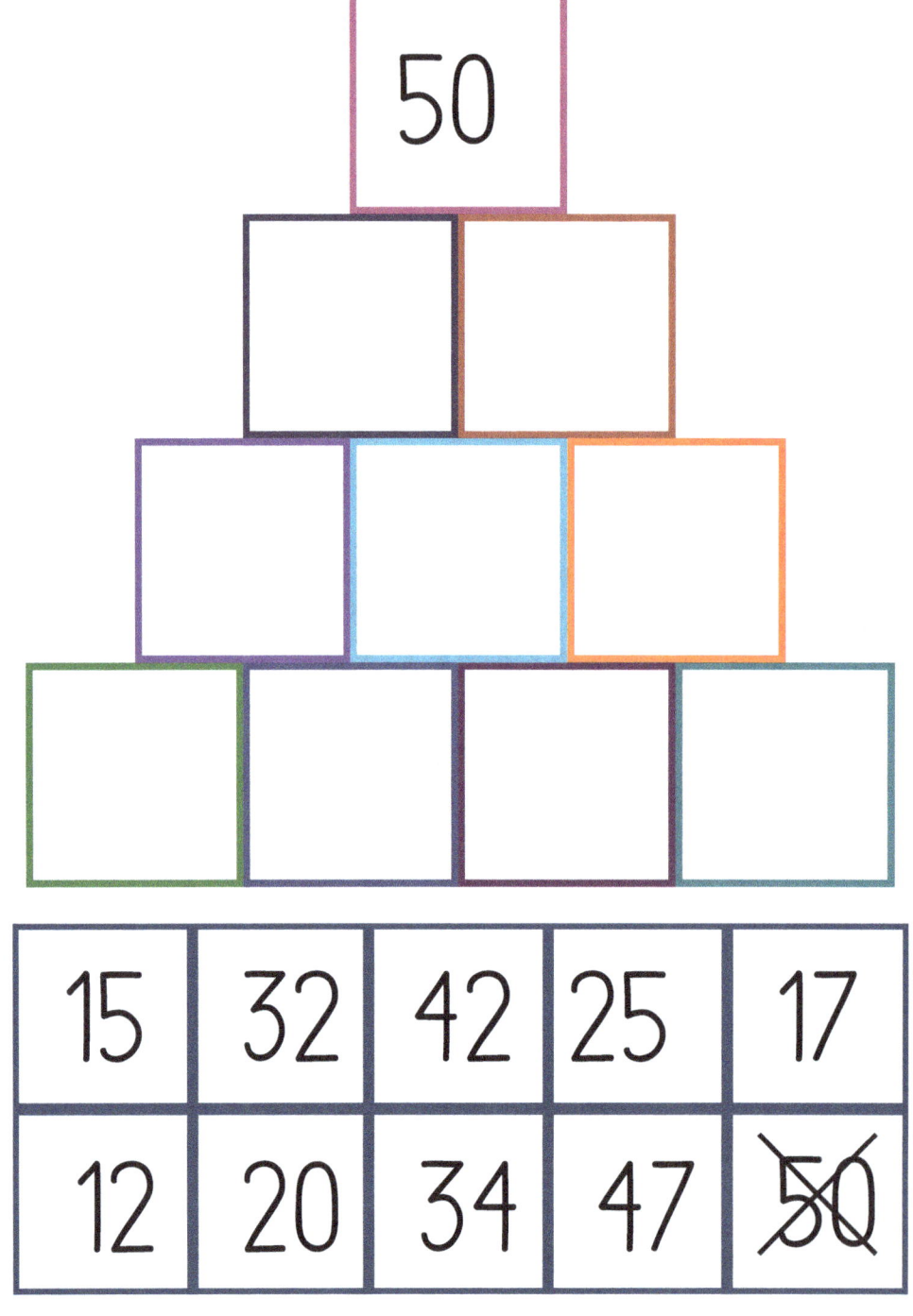

50				

15	32	42	25	17
12	20	34	47	~~50~~

PYRAMIDS OF NUMBERS

Order the pyramid from largest to smallest with the indicated numbers

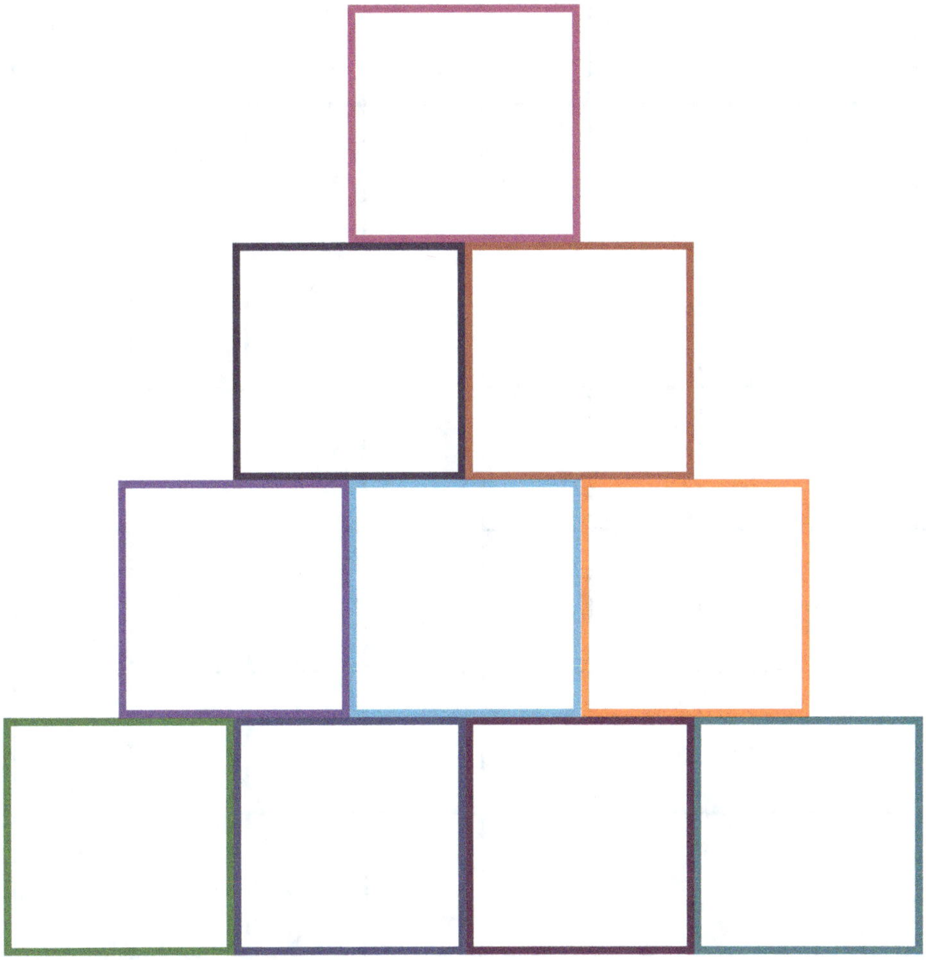

35	44	19	27	18
45	28	43	31	49

LET'S COUNT TO 100

Fill in the missing numbers to reach 100

									10
									20
									30
									40
									50
									60
									70
									80
									90
									100

PYRAMIDS OF NUMBERS

Order the pyramid from largest to smallest with the indicated numbers

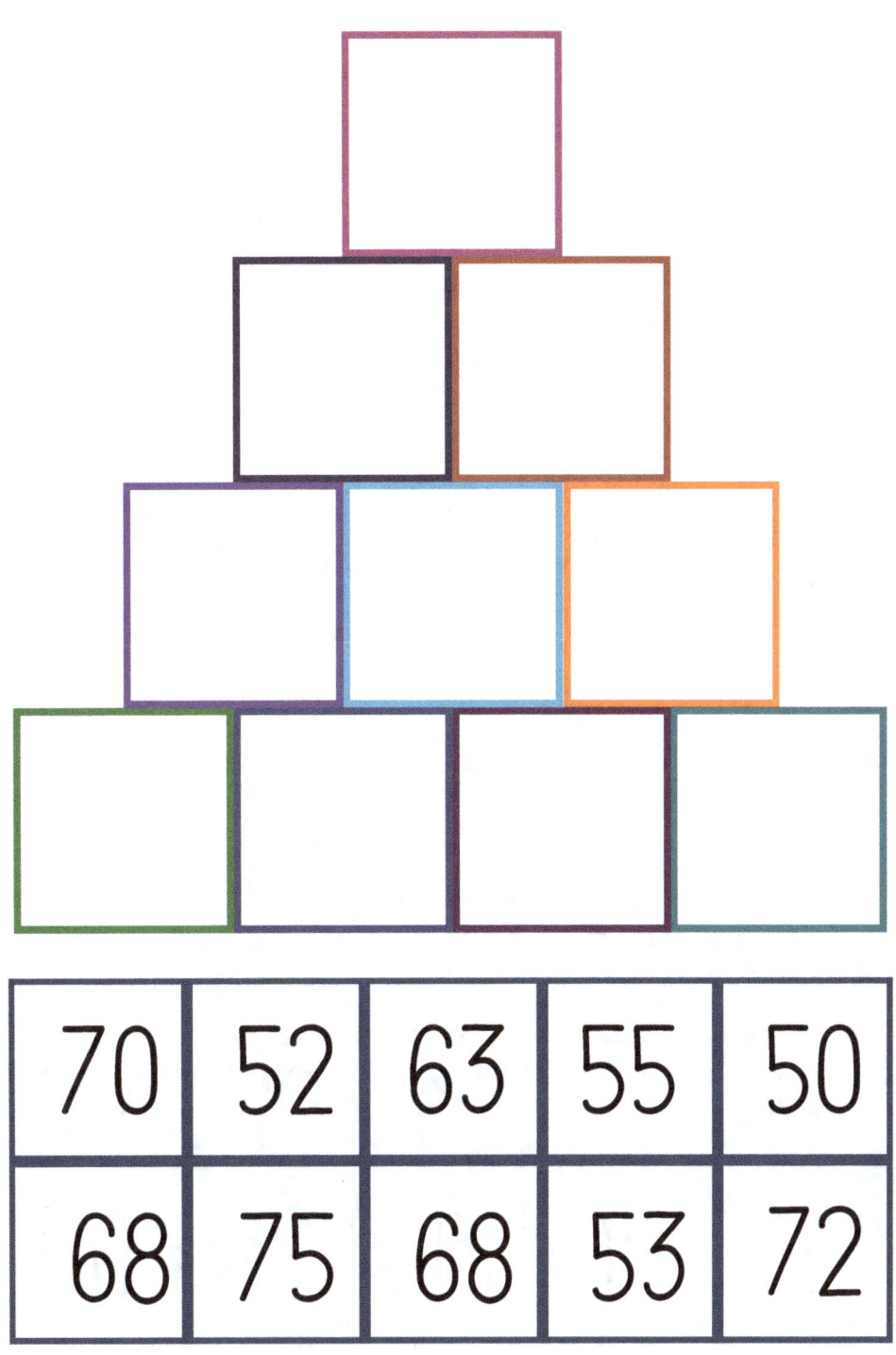

70	52	63	55	50
68	75	68	53	72

PYRAMIDS OF NUMBERS

Order the pyramid from largest to smallest with the indicated numbers

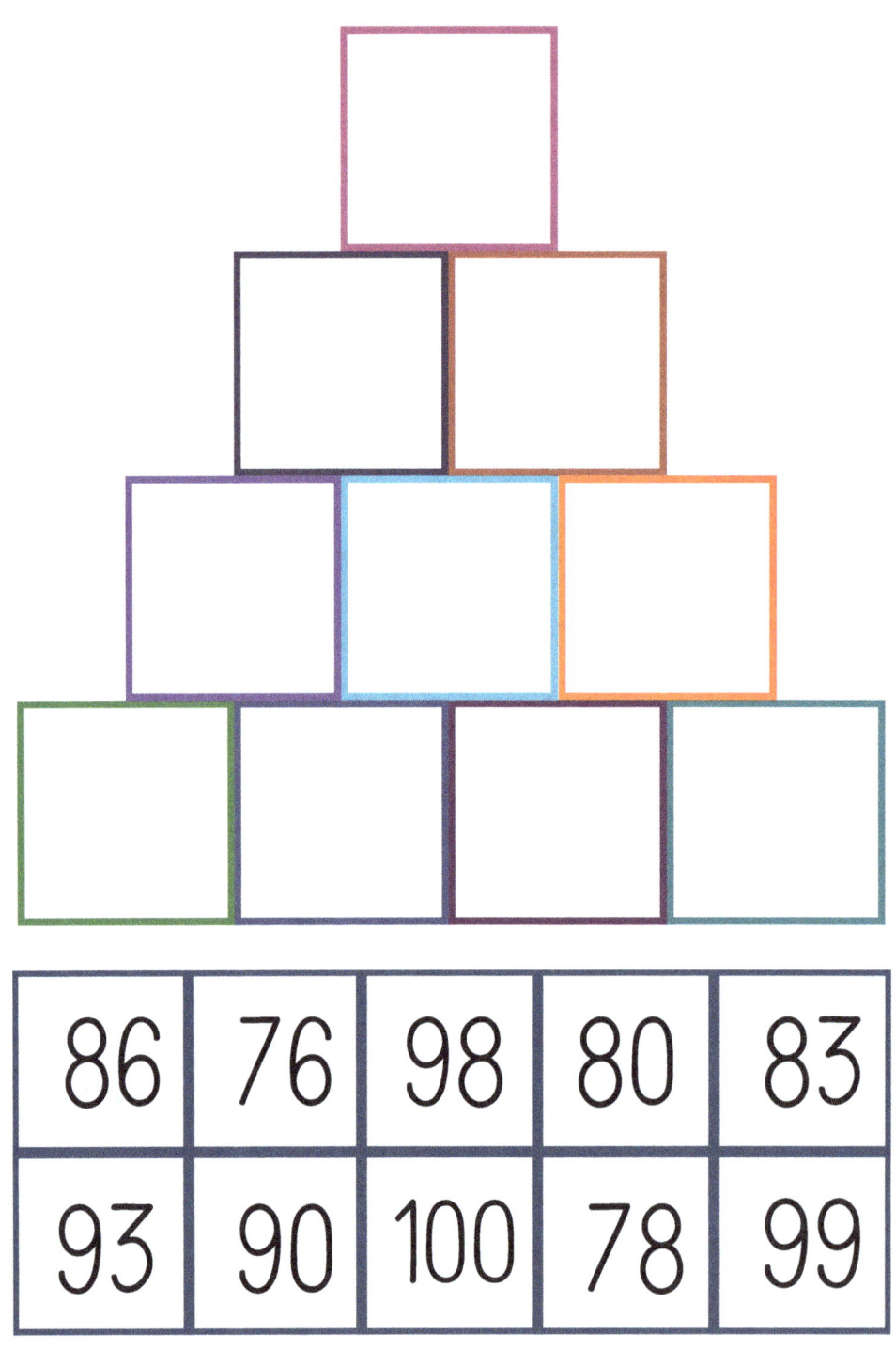

86	76	98	80	83
93	90	100	78	99

NUMERICAL ORDER

Order the numbers from least to greatest

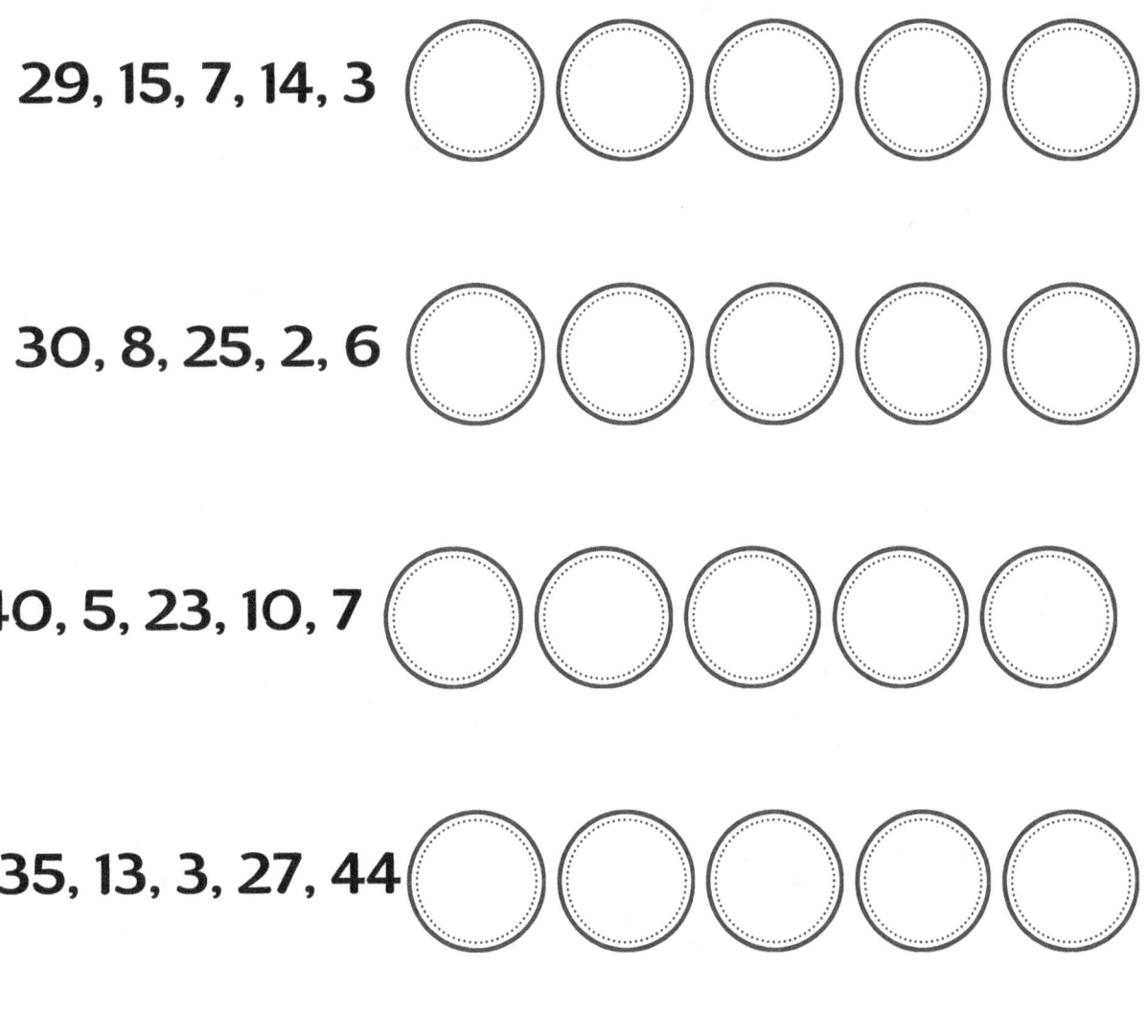

29, 15, 7, 14, 3

30, 8, 25, 2, 6

40, 5, 23, 10, 7

35, 13, 3, 27, 44

33, 1, 48, 39, 10

NUMERICAL ORDER

Order the numbers from least to greatest

12, 5, 7, 20, 1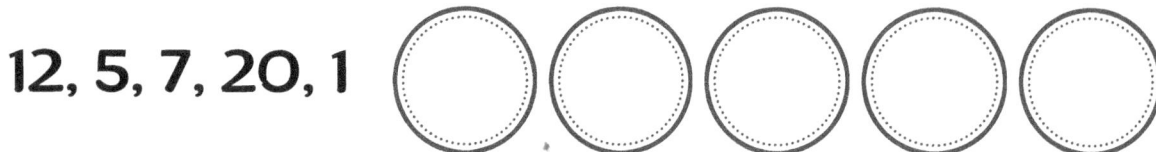

9, 15, 12, 4, 20

11, 8, 2, 6, 25

3, 8, 13, 25, 17

19, 10, 3, 22, 14

NUMERICAL ORDER

Order the numbers from greatest to least

12, 5, 7, 20, 1

9, 15, 12, 4, 20

11, 8, 2, 6, 25

3, 8, 13, 25, 17

19, 10, 3, 22, 14

NUMERICAL ORDER

Order the numbers from greatest to least

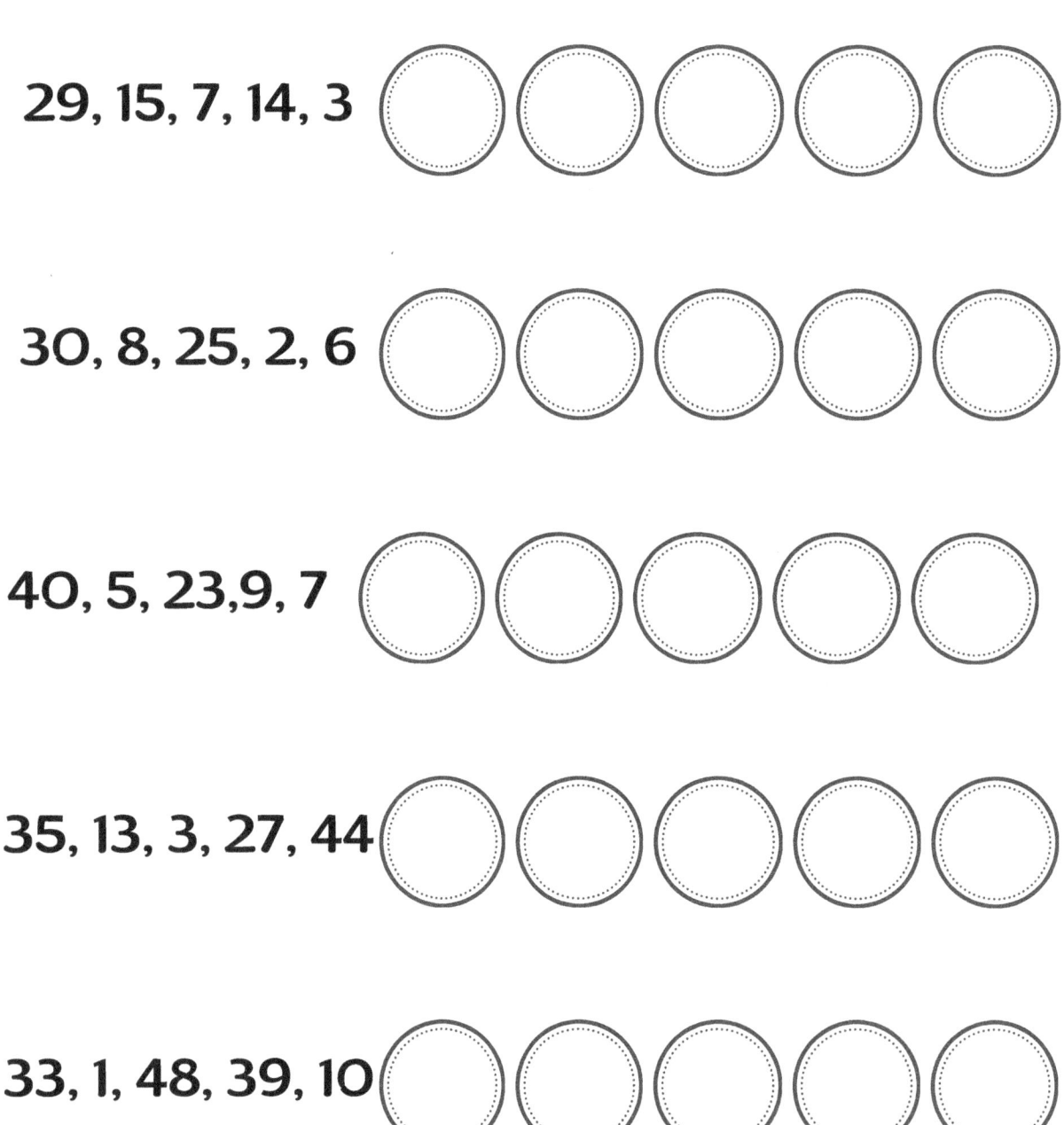

29, 15, 7, 14, 3

30, 8, 25, 2, 6

40, 5, 23, 9, 7

35, 13, 3, 27, 44

33, 1, 48, 39, 10

JUMP COUNTING

Fill in the missing numbers

1 in 1 account

24

28

30

Count by 2 by 2

30

38

44

JUMP COUNTING

Fill in the missing numbers

Count by 3 by 3

46

52

67

Count by 4 by 4

62

70

78

90

JUMP COUNTING

Fill in the missing numbers

Count by 5 by 5

15

20

40

Count by 6 by 6

6

24

42

DISCOVER THE PATTERN

Write the correct numbers in each circle

DISCOVER THE PATTERN

Write the correct numbers in each circle

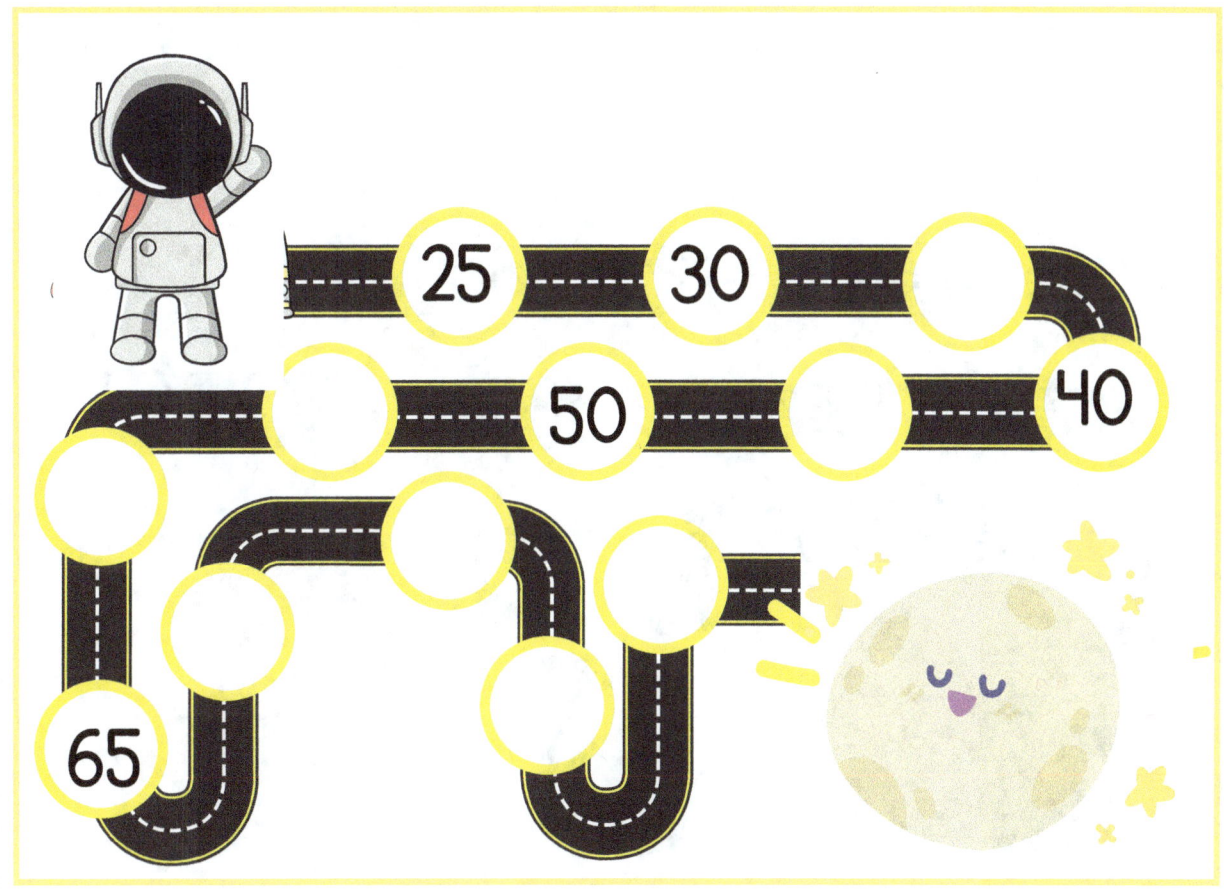

DISCOVER THE PATTERN

Write the correct numbers in each circle

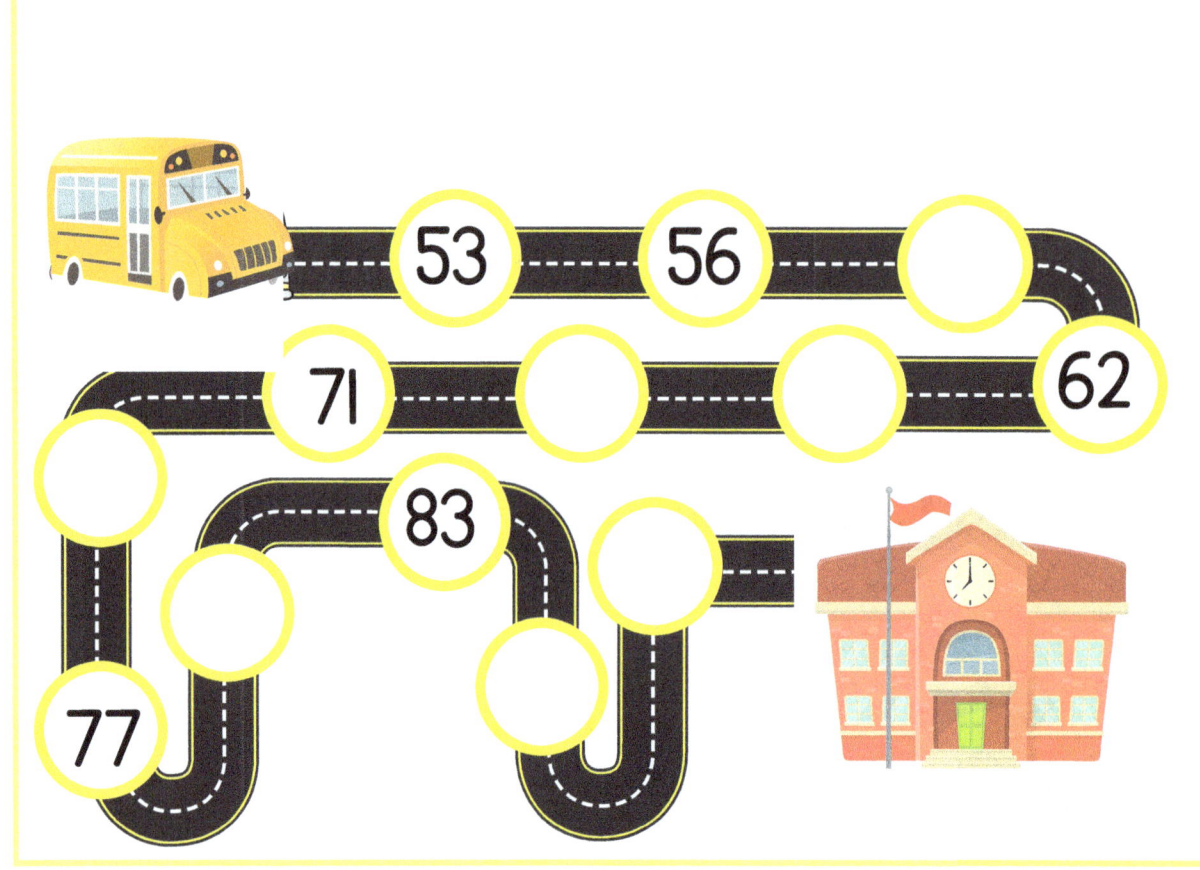

DISCOVER THE PATTERN

Write the correct numbers in each circle

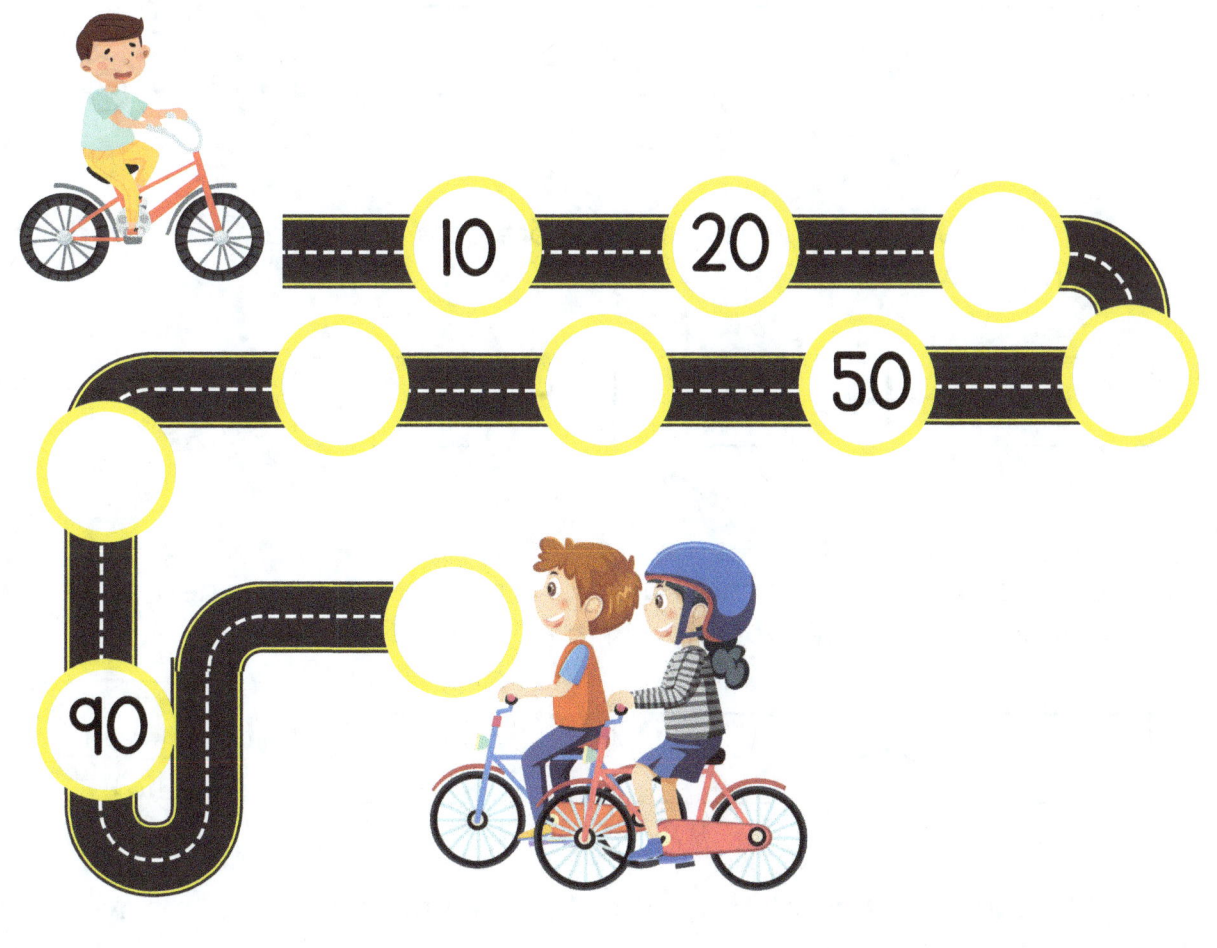

NUMERICAL CHALLENGES

Solve the following math problems

- There are 12 cars in the parking lot and 5 more arrive. How many are there in total?

  [ ] **+** [ ] **=** [ ]

- A train has 9 cars and they add 8 more cars. How many cars does the train now have?

  [ ] **+** [ ] **=** [ ]

- There are 15 passengers on a bus and 3 more get on at the next stop. How many passengers are there now in total?

  [ ] **+** [ ] **=** [ ]

- A ship carries 10 boxes of toys and 13 boxes of candy. How many boxes does it carry in total?

  [ ] **+** [ ] **=** [ ]

NUMERICAL CHALLENGES

Solve the following math problems

- A train leaves the station with 48 passengers. At the next station 14 passengers get off. How many passengers were left on the train?

$$\boxed{} - \boxed{} = \boxed{}$$

- 27 children participated in a bicycle race, 15 reached the finish line. How many children were left to reach the finish line?

$$\boxed{} - \boxed{} = \boxed{}$$

- A plane has 65 seats in total, 41 are occupied. How many unoccupied seats are there?

$$\boxed{} - \boxed{} = \boxed{}$$

- Lie has 35 toy cars and gives 13 to Leslie. How many cars does Lie keep?

$$\boxed{} - \boxed{} = \boxed{}$$

LET'S ADD WITH IMAGES

Observe carefully and solve the sums

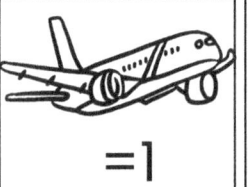 =1  =3 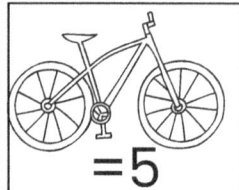 =5  =7  =9

 +  =

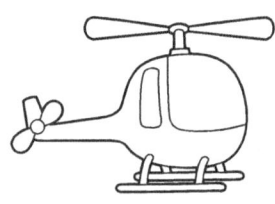 +  =

 +  =

 +  =

LET'S ADD WITH IMAGES

Observe carefully, remember the value and solve

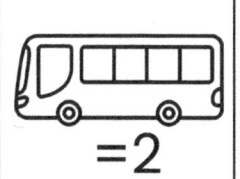 =2   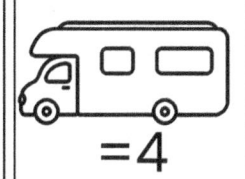 =4   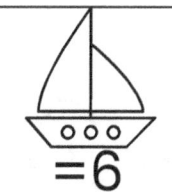 =6   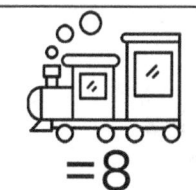 =8    =10

 +  =

 +  =

 +  =

 +  =

DISCOVER THE VALUE

Observe each operation and discover the value of each image. Reply

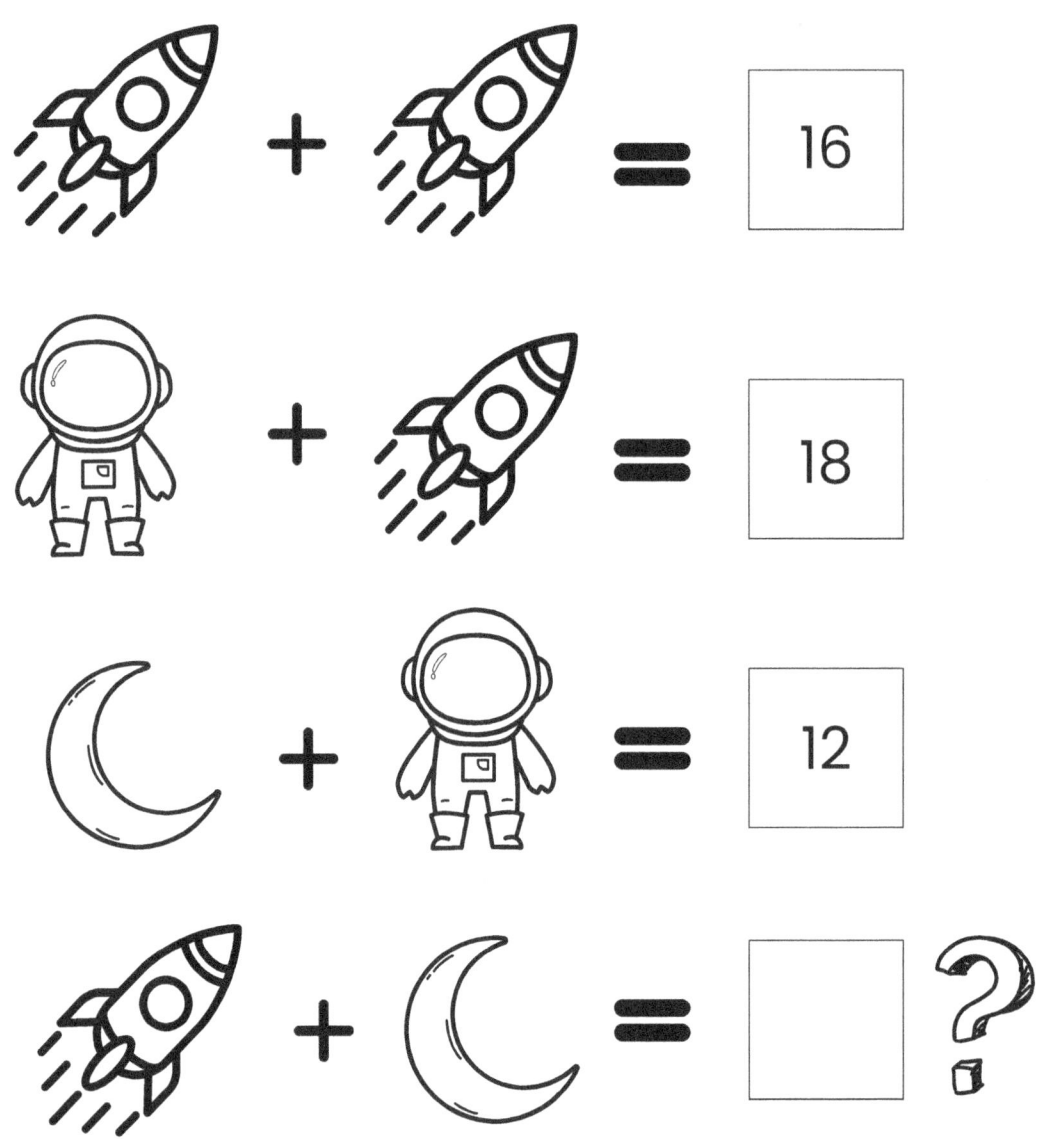

DISCOVER THE VALUE

Observe each operation and discover the value of each image. Reply

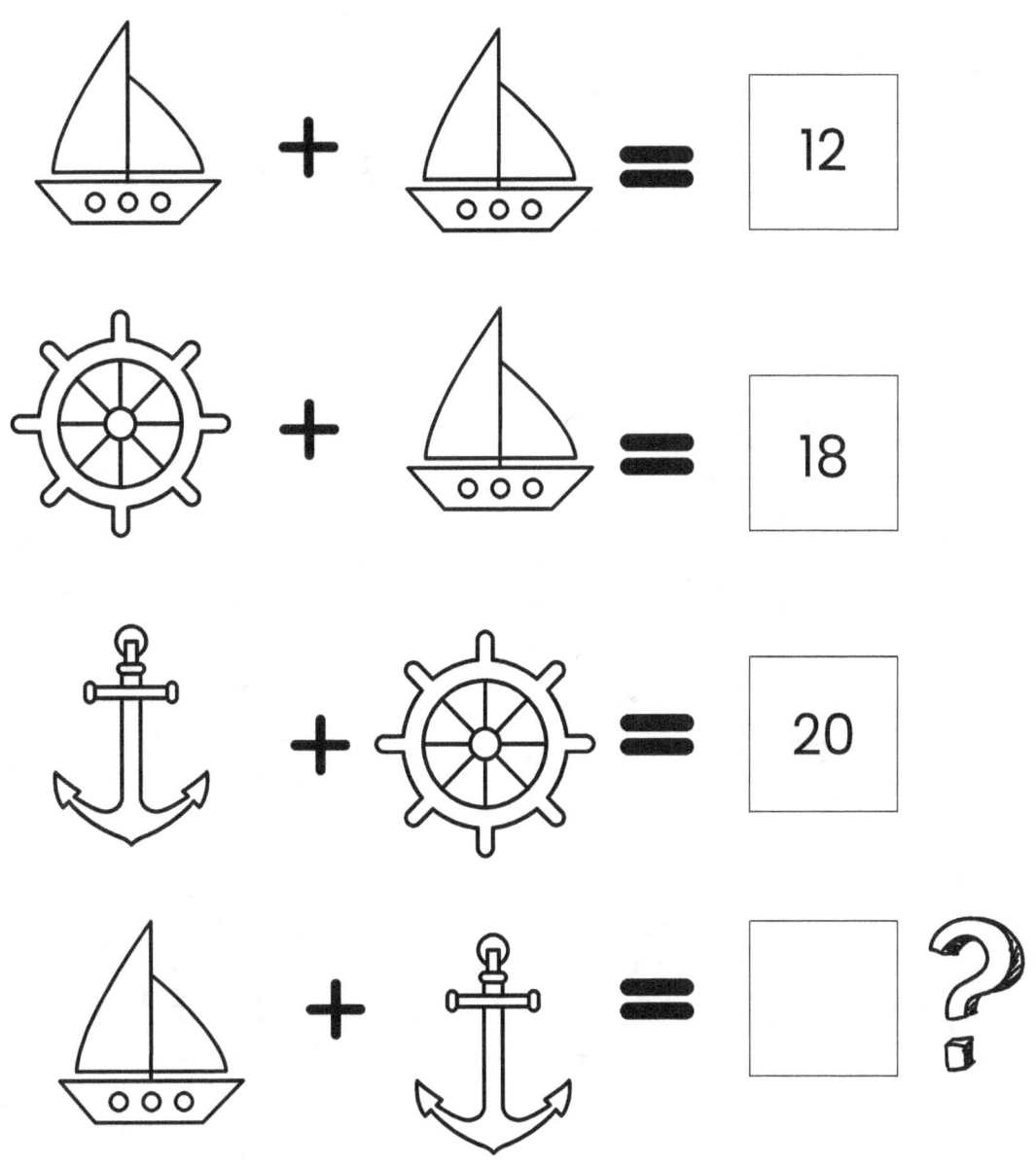

DISCOVER THE VALUE

Observe each operation and discover the value of each image. Reply

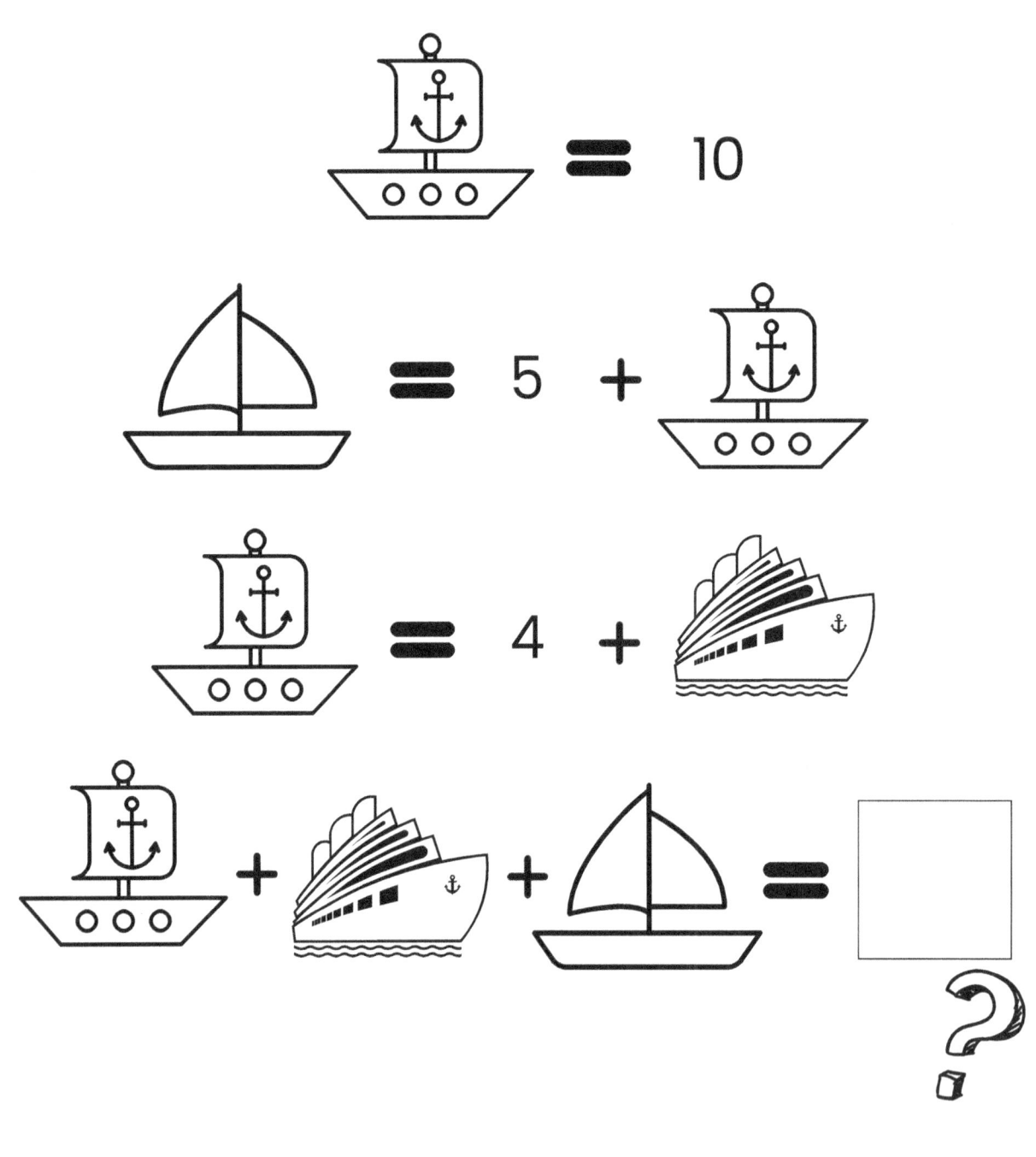

DISCOVER THE VALUE

Observe each operation and discover the value of each image. Reply

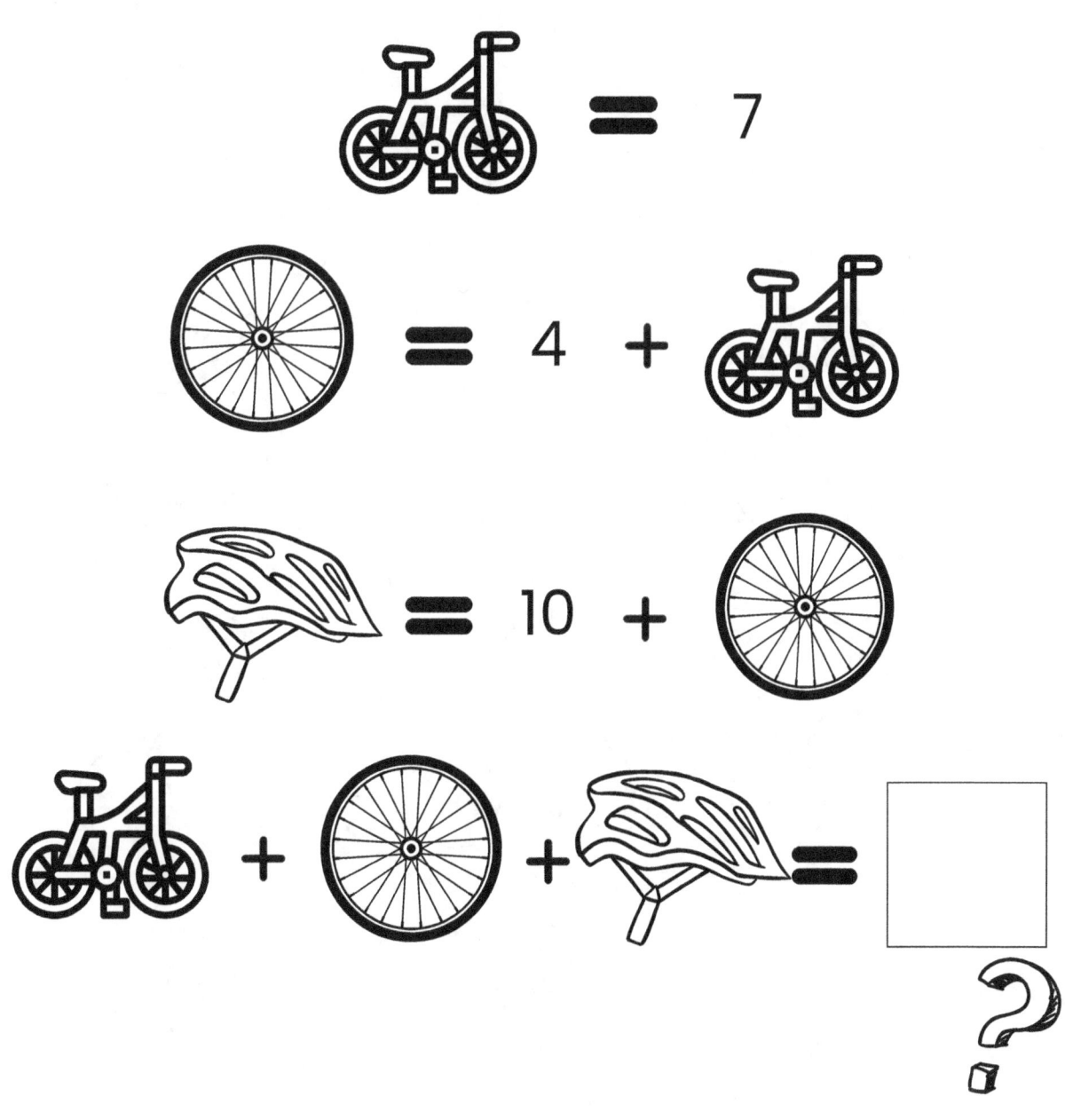

LOST NUMBERS

Observe carefully, taking into account the "+, -" signs to solve the puzzle

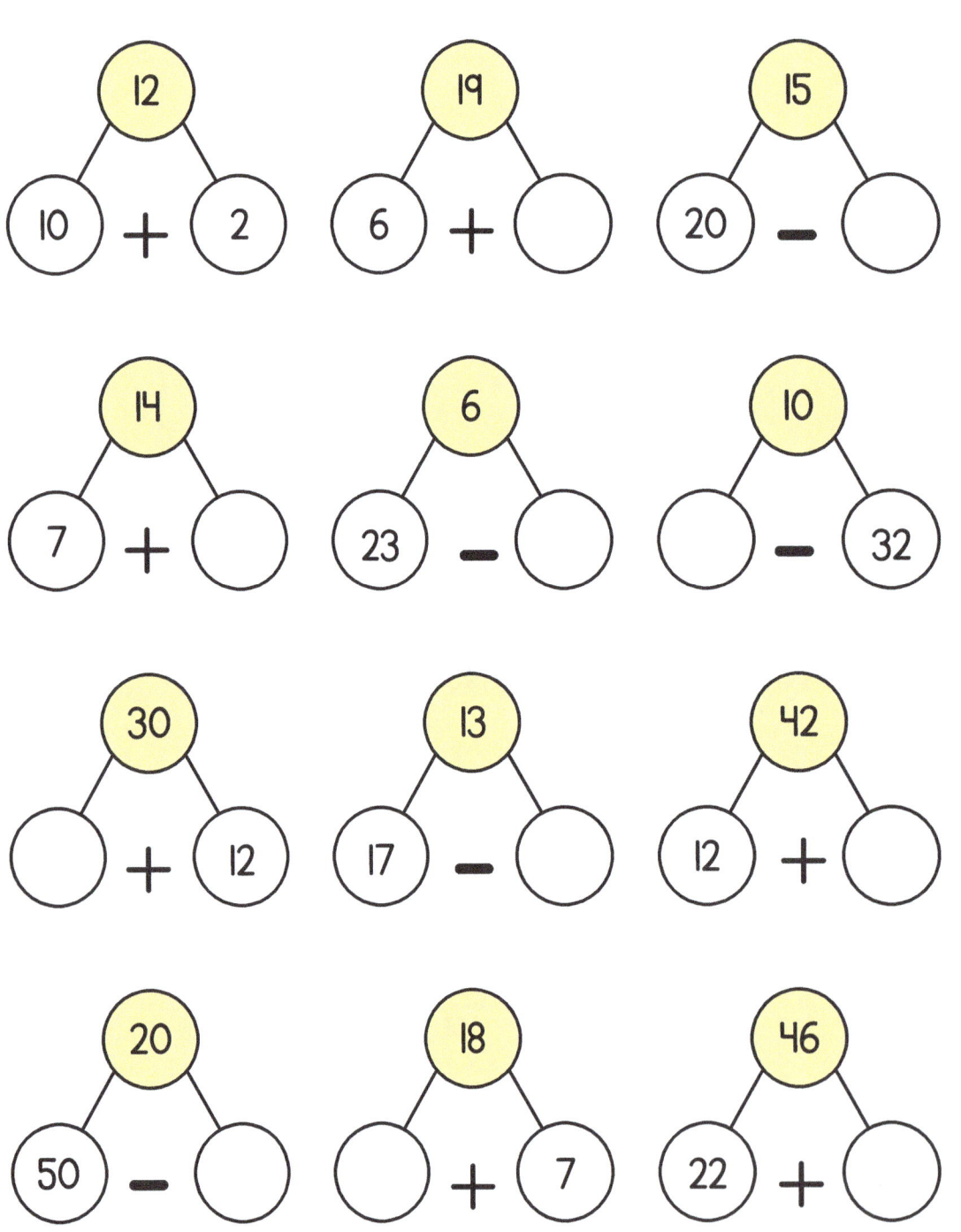

LOST NUMBERS

Observe carefully, taking into account the "+, -" signs to solve the puzzle

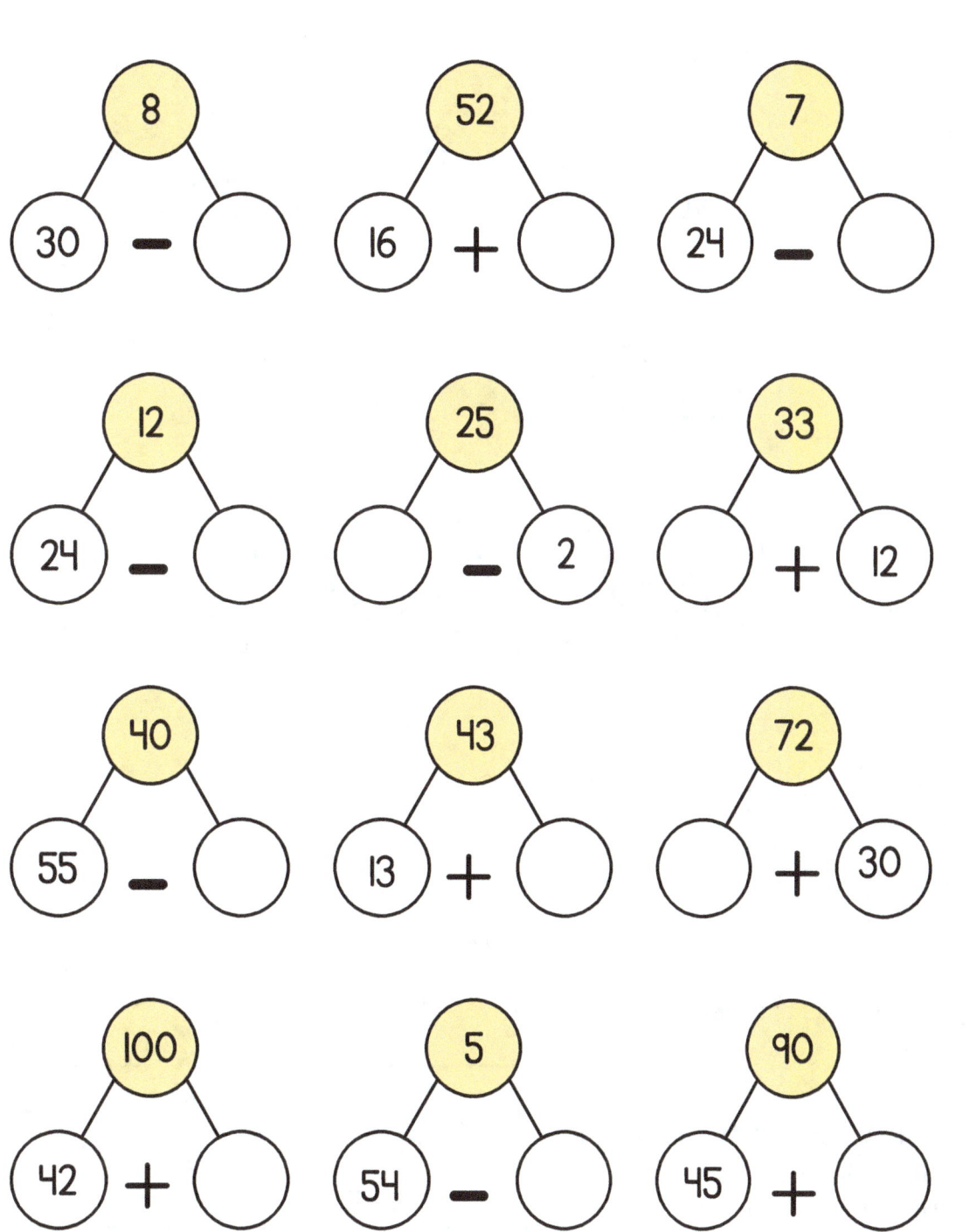

LETS GO SHOPPING!

Look at the prices and respond

- What is the most expensive toy? _____

- What is the cheapest toy?

_____

- Which toys total $40?

_____

- Which toy costs more than the helicopter but less than the rocket?

_____

LETS GO SHOPPING!

Look at the prices and respond

You went to the bakery with $42, select the breads you want to buy and that you can afford (circle them), write the prices on the receipt and add up.

$12 $15

$8 $10

Bakery

$18 $9

Your change is: _____

SUDOKUS

Sudoku 1

2			
	4	2	
3			2
4		3	

Sudoku 2

3			1
	3	1	4
	1		2

SUDOKUS

Sudoku 3

		3	2
3	2		
	1		
4		2	

Sudoku 4

			2
4	2		1
2	3		
			3

SUDOKUS

Sudoku 5

2		1	6	3	4
	3			2	
	1	3			2
	6			1	
1		6	3		5
3	2	5			6

SUDOKUS

Sudoku 6

1		4		3	2
		2			
3				6	1
		1		5	
	2	3	4	1	6
		6			3

SUDOKUS

Sudoku 7

2		1	6		4
	3				
	1	3	5		2
		2	4	1	
1		6	3		5
3		5			6

SUDOKUS

Sudoku 8

		3			5
	5	4	6	3	
	3			1	
2		1	4		3
		6	5		4
5		2	3		

CERTIFICATE

It is hereby certified that:

....

....

He successfully completed this book and demonstrates
mastery of the subject.

....

Mom/Dad/Te
acher

Student

You did!

Mariledys T.
Author

May you be successful in all your academic
activities and achieve your goals!

SUDOKU SOLUTION

Sudoku 1

2	3	1	4
1	4	2	3
3	1	4	2
4	2	3	1

Sudoku 2

3	4	2	1
1	2	4	3
2	3	1	4
4	1	3	2

Sudoku 3

1	4	3	2
3	2	1	4
2	1	4	3
4	3	2	1

Sudoku 4

3	1	4	2
4	2	3	1
2	3	1	4
1	4	2	3

SUDOKU SOLUTION

Sudoku 5

2	5	1	6	3	4
6	3	4	2	5	1
4	1	3	5	6	2
5	6	2	4	1	3
1	4	6	3	2	5
3	2	5	1	4	6

Sudoku 6

1	5	4	6	3	2
6	3	2	1	4	5
3	4	5	2	6	1
2	6	1	3	5	4
5	2	3	4	1	6
4	1	6	5	2	3

Sudoku 7

2	5	1	6	3	4
6	3	4	2	5	1
4	1	3	5	6	2
5	6	2	4	1	3
1	2	6	3	4	5
3	4	5	1	2	6

Sudoku 8

6	2	3	1	4	5
1	5	4	6	3	2
4	3	5	2	1	6
2	6	1	4	5	3
3	1	6	5	2	4
5	4	2	3	6	1

Dear readers,

Thank you for embarking on this exciting mathematical logic adventure with me! It has been a pleasure to create this workbook designed especially for curious and creative minds.

Support from readers like you is invaluable to indie authors like me. If you have a moment, I would greatly appreciate it if you could leave a review on Amazon. Your opinions are crucial so that more parents, teachers or adults who love early childhood education find these exercises.

I invite you to visit my page on Amazon and by following me you can access the different books that I am preparing for our children.

Thank you for contributing to the reading community and making it possible for more people to discover this book!

With gratitude, Mariledys

Scan to leave your comment or visit my page on Amazon